THE SELFISH APE

THE SELFISH APE

Human Nature and Our Path to Extinction

NICHOLAS P. MONEY

REAKTION BOOKS

Published by
REAKTION BOOKS LTD
Unit 32, Waterside
44–48 Wharf Road
London N1 7UX, UK
www.reaktionbooks.co.uk

First published 2019

Printed and bound in Malta
by Gutenberg Press Ltd

A catalogue record for this book is available from the British Library

ISBN 978 1 78914 155 9

CONTENTS

PREFACE

> IS IT, indeed, true, that the Poet, or the Philosopher, or the Artist whose genius is the glory of his age, is degraded from his high estate by the undoubted historical probability, not to say certainty, that he is the descendent of some naked and bestial savage, whose intelligence was just sufficient to make him a little more cunning than the Fox, and by so much more dangerous than the Tiger?
>
> THOMAS HUXLEY, *Man's Place in Nature* (1863)

This book is about what we are. Looking in the bathroom mirror in the morning, I have been struck, from time to time, by the foolishness of the animal blinking back at me. When the prospects of the day seem unusually bright, the mirror frames the Laughing Cavalier, but a rather melancholic creature appears more often. Whatever the reflection, the time spent pulling at my skin and putting the fear of imminent demise to sleep for another day suggests some level of vanity. In our age of extreme self-absorption, my case of egotism is comparatively mild, but I did write a song some time ago that argues differently. It is suited for a Victorian music hall and the high-pitched voice of an overly sensitive youth. Here is the first verse:

How cruel is life, to me, today,
That I'm not rich and famous?
And I am forced to just scrape by,
And try to be courageous.

But enough about me. All of us belong to a species of African ape to which Carl Linnaeus applied the Latin name *Homo sapiens*, or wise man, in 1758. He must have felt very confident about our brilliance at that time. Fantastical thinking throughout human history has led us to the strangest delusions about our significance in nature and there is a dreadful persistence to claims that we are better than the rest of the biological world and that we are forging a brighter future through our technological brilliance. According to the view of one popular thinker, we are already assuming the role of a new version of man with godlike powers: *Homo deus*.[1] With our collective wisdom in such short supply in this twenty-first century, and the international energy invested in navel-gazing, *Homo egotisticus* seems more appropriate; or, better still, *Homo narcissus* – self-absorbed man.

The outline of the Narcissus myth is probably familiar to readers, but a refresher may be helpful. As the Roman poet Ovid tells it in his *Metamorphoses*, Narcissus was the beautiful teenage son of a river nymph called Liriope. Girls and boys, as well as spirits of the woods and waters, were captivated by the young man. Narcissus enjoyed the attention, but he spurned every attempt at intimacy. One of the rejected male admirers prayed that Narcissus would be punished with a taste of his own medicine, and this wish was granted in spades by the goddess Nemesis. Seeking a restful spot in the woods, Narcissus became spellbound by his reflection in a pool of clear water. Falling deeply in love, and exasperated that he could not embrace this dazzling young man, he realized after a while that the object of his desire was himself. Rather than bringing him back to his senses, this insight only deepened his desire. Anguished beyond bearing, he willed himself to death.

Before we allow ourselves to feel superior to the poor lad, consider that his triumph of personal absorption over self-preservation applies to humanity today as we demonstrate our inability, or unwillingness, to wrestle with climate change. Ours is an expression of narcissism beyond the dreams of Ovid. We are cosmic vandals. In the eighteenth century, Edward Gibbon wrote with brilliance about *The History of the Decline and Fall of the Roman Empire*; there will be no historian to write *The History of the Decline and Fall of the Earth*. Three centuries after Linnaeus, we have all the evidence required for a rechristening:

> *Homo narcissus: illa simiae species Africana ab origine quae adeo orbem pervastavit terrarum ut ipsa extincta fiat.*[2]

> *Homo narcissus*: species of ape of African origin that devastated Earth's biosphere and thereby drove its own extinction.

We would be well advised to take a more objective look at ourselves, to appreciate what we are and what we are not. This short book is designed as a recalibration device. We begin with our location in the cosmos (Chapter One), our microbiological origins, how our bodies function and how we are encoded in DNA (Chapters Two to Four). This leads to an exploration of human reproduction, brain function and ageing and dying (Chapters Five to Seven). Chapters Eight and Nine concern the intertwined phenomena of human greatness and failure; intellectual greatness has come from experimental science, but progress in understanding and manipulating nature has come at the expense of destroying the surface of Earth. By any measure, we have behaved rather badly. Chapter Ten considers the fate of civilization with the hope that by facing the truth we will transcend our self-absorption and elevate *Homo narcissus*, providing some redemption and justifying the name *Homo sapiens*.

With such magnificent brains, it is tempting to believe that even the biggest problems can be fixed, that technology will save us, that Chicken Licken was wrong. P. G. Wodehouse framed the fluffiness of this thinking very gently:

> I'm not absolutely certain of my facts, but I rather fancy it's Shakespeare . . . who says that it's always just when a chappie is feeling particularly top-hole, and more than usually braced with things in general that Fate sneaks up behind him with a bit of lead piping.[3]

The clock is ticking. The Four Horsemen will be here in the blink of a geological eye.

ONE

GLOBE

How Life Lends Itself to Earth

We spend most of our lives on the surface of the Earth, in contact with the ground and breathing in the atmosphere. We walk, run, sit and sleep on land. We inhale and exhale a mixture of gases from first breath to last gasp. All of our companions, from the largest whales to the smallest viruses, exist in the 20-km-thick (12-mi.) skin of Earth's biosphere.[1] High above, in the upper atmosphere, even the most resilient organisms dry out and are burned to crisps by the Sun. Below the biosphere, in the depths of the crust, life is extinguished by heat radiating from the underlying mantle.

Oodles of physical peculiarities support the biological activity here. Earth traces a Goldilocks orbit, at a distance from the Sun which allows water to be liquid: not too close to make it boil away and not too far to make it freeze. The Sun is a medium-sized, middle-aged star, classified as a yellow dwarf by cosmologists. Yellow dwarfs are nuclear reactors that fuse atoms of hydrogen into helium and release a lot of energy. Our star is 4.6 billion years old. It will continue to burn for another 5 billion years before it runs out of its hydrogen fuel and swells into a weaker type of star called a red giant. Long before then, in a billion years or so, the ageing Sun will turn brighter and its awful incandescence will sterilize the biosphere for good.

What good fortune, then, to be born when the Sun is glowing just right. Also, because our galaxy – the Milky Way – is almost as old as the universe, it contains the chemicals needed for life. Atoms of carbon that form the skeletons of our proteins, and other organic molecules, could not have formed until the first stars created after the Big Bang detonated to form supernovae. Three billion years into the history of the universe, these firework displays began recycling stardust to fertilize the next generation of suns that contained the heavier elements. The reason that there is plenty of carbon and other 'metals' to go around now is that the galaxy has entertained multiple cycles of collapsing and exploding stars to flood itself with these elements.[2]

Without the Sun behaving as it does, and without the galaxy being as old as it is to have formulated the chemical ingredients for building living things, we would not be here. Delving further into the workings of physics and chemistry, some scientists argue that the universe is fine-tuned in a fashion that supports life. The force of gravity is an example of one of these fortunate characteristics. If gravity were slightly weaker, matter would never become compressed into stars in the first place. Conversely, a stronger gravitational force would have prevented the expansion of the universe and ended the festivities in a Big Crunch soon after the Big Bang.

These musings about the fortunes of the physical world are weakened when we see that they depend upon a circular argument. Rather than believing that the universe is organized for our benefit, it makes much more sense to consider the ways that biology has fitted itself to the available circumstances. Every feature of every animal, plant and microbe is superbly adapted for living on this planet. And, in support of this interpretation, we have known exactly how this has been brought about for more than 150 years – ever since Charles

Darwin explained the mechanism of natural selection. Evolution does not require any circular reasoning to make sense.

The mechanism of evolution is so pervasive that it may whip up life of some kind on every Goldilocks planet. The anthropic principle, which is related to the fine-tuning argument, suggests that the universe has to be compatible with some form of consciousness and could not exist without someone appreciating it. This is another circular argument that is as impossible to refute as it is to take seriously. Consciousness among animals like us is a product of evolution. We may regard it as a fortunate characteristic, but it does not take much imagination to view it instead as a widespread curse. Would the prisoner, aware that something unpleasant is about to happen in the dungeon, prefer to be blessed by obliviousness?[3] And as for not looking life and other gift horses in the mouth, it is useful to remember that nobody ever asked to be born. Indeed, some modern philosophers argue that the worst possible thing that anyone can do is to become a biological parent. The problem here, and it is a big one, is that the birth of ever more beings with the capacity for suffering adds to the collective horror in the universe.[4] This psychic concern overlaps with the more practical issue of the environmental damage caused by billions of humans.

Putting aside the questionable virtues of parenthood, the notion of a universe privileged for us and by us exposes the astonishing arrogance of our species. With or without humans, Earth will spin about its polar axis at 1,600 km (1,000 mi.) per hour and fly around the Sun at 108,000 km (67,000 mi.) per hour, while the entire solar system sweeps around the centre of the Milky Way.[5] All this orbital motion originated in clouds of interstellar dust and gas that became spotted with concentrations of mass. As more and more material swirled towards these nuggets under the pull of gravity, islands of density grew into new stars. Each star is accompanied by circulating planets and each planet spins around its own axis. Planets are the remnants of the disc of dense gas from which their sun was born; they continue to orbit their stars and entire galaxies whizz around at great speeds because their motion is unopposed by anything in space.

And here we walk, run, sit and sleep, on the third planet from our star, on the Orion Arm of the Milky Way galaxy, in the middle of the observable universe. There is nothing special about our apparent position. It is just that we can only peer so far in any direction and by sight or radio telescope any resident will always find itself in the dead centre of a sphere. Think about being at sea in a kayak, too far from land to glimpse any coastlines. You can paddle around a good deal out there, while seeming to remain in the middle of a big circle. The circle of sea and the sphere of the universe move with the observer. It is, however, always possible that the Milky Way is close to one end of an egg-shaped universe. We would not know otherwise.

If we contemplated our cosmic placement more frequently, I wonder if we might feel agoraphobic? Or, perhaps claustrophobia could be a more natural cause of a species-wide panic attack? Stephen Hawking seemed influenced by claustrophobia when he recommended that we work on an interstellar escape plan with some urgency.[6] Unfortunately, he did not suggest how we might propel ourselves over trillions of kilometres of space without being blown to smithereens by radiation from supernovae. Really supportive universes would have less in the vein of cosmic rays, and restaurants for interstellar astronauts would be a nice touch too. As things stand, it is *almost* as if we are not the intended beneficiaries of all this creative work on the part of gravity.

Until science began to displace the classical cosmology of Aristotle, we imagined that the stars were painted on a crystal sphere, bleached by sunbeams during the day and revealed as lesser lights when the Sun dropped below the horizon. We might have supposed that this decorative vault was quite close to us, not far above the clouds. Hamlet considered the 'brave o'erhanging firmament' as 'a foul and pestilent congregation of vapours' (Act II, Scene 2), whereas Milton rejoiced in the galaxy, 'Which nightly as a circling zone thou seest/ Powdered with stars' (*Paradise Lost*, Book VII, 580–81). The slow transit of a bright comet was a cause for misgivings, with its fiery tail streaking the roof above the stationary Earth. There was certainly a lot going on up there. Some of the stars showed up in the same position relative to one another, while others moved from place to place each night: Mercury, Venus, Mars, Jupiter and Saturn – Milton's 'five other wandering Fires, that move/ In mystic dance' (*Paradise Lost*, Book V, 177–8). Everything seemed to be arranged for us and forces that we could not hope to understand animated this clockwork sky. We were, at once, subjugated by gods and empowered by their interest in everything we did.

Humanity began to cross 'the vast gulf of the monkish and deluded past' towards the modern era of the objective exploration of nature in the seventeenth century.[7] Cosmology became a subject of intense scientific inquiry, bookended by Galileo's *Dialogue Concerning the Two Chief World Systems*, published in 1632, and Isaac Newton's *Principia* of 1687. Galileo made an impassioned case for the revolution of the Earth around the Sun, rather than vice versa, and Newton derived the laws of motion and gravity that maintained the orbits of planets. Four centuries on, we have a firm grasp on the physics of the universe in the aftermath of the Big Bang. The workings of matter in the instant of the beginning of time, called Planck time, are baffling, but we have certainly come a long way.[8] Far enough, I think, for most of us to make sense of life without worrying about the details of the birth of the universe. It is here and we live in it.

We have established that Earth is a good place to live – or would be, without all the mistakes made by humans. Environmental conditions vary a good deal on top of the crust. Seventy-one per cent of the surface is submerged by salty water. Most of the rest of the real estate is above water and is greened by forests and grasslands, or browned and yellowed as deserts. We do not thrive in polar climates, or in hot desert temperatures much above 120 degrees Fahrenheit (approaching 50 degrees Celsius). Continuous hydration can keep the fittest going in places like Death Valley, in California, during a day hike, but this environment tests the limits of human resilience. Ultraviolet rays from the Sun are another hazard, and we rely on a 3-mm (0.12-in.) sliver of ozone in the stratosphere for protection. Without this benevolent gas, the DNA in our skin would be scrambled beyond repair unless we sought refuge in caves. The presence of ozone might be viewed as a further example of the fine-tuning of this best of all possible worlds.[9] Sidestepping wishful thinking, the scientific truth is that the ozone layer was here and we evolved under it, becoming as tolerant of the incoming radiation as we needed to be – at least before we weakened the shield with refrigerants – and not a jot more.

Biology happens in biomes, which are categories of vegetation and associated wildlife. Ecologists recognize more than a dozen kinds of biome, including tropical broadleaf forests, temperate grasslands and mangrove swamps. A good deal of Earth's natural vegetation has been supplanted by cereal agriculture, which tends to work best in places that had once supported an abundance of wildlife. Major cities have flourished in natural oases too, although plenty of humans live in hot deserts where fresh water is furnished by irrigation and desalination.

Our well-being depends on access to clean – or at least clean-ish – water and air, and to a variety of fruits and vegetables. The consumption of animals has been an invariable feature of human history, but we can adopt a vegetarian diet if no meat is available, or for ethical, economic and environmental reasons. With or without meat, we are nothing without botany. Plants are so overwhelmingly important in human affairs that their study deserves the reverence afforded to business and accountancy in modern colleges and universities. The intellectual underpinning of a business degree is gossamer thin. 'Knowledge is Power' is chiselled on the stone gateway leading to the School of Business at my university. The phrase is attributed to Thomas Hobbes, the political philosopher, and first appeared in the 1668 Latin edition of his *Leviathan*, as *scientia potentia est*.[10] In this great work, Hobbes suggested that the importance of science, or objective knowledge, lies in its practical applications. He would chuckle at the association of his aphorism with the rather miserable aspirations of investment bankers.

In any case, educated citizens of the twenty-first century should have some appreciation of the botanical foundations of our existence. Everyone should be able to take a stab at explaining where food comes from, and the correct answer should not stop with 'grocery store' or 'supermarket'. The process begins with entropy and ends with sugar. Entropy is the term for the physical process that makes a mess of everything. It applies to the transformation of a library with books stacked on shelves into a pile of rubble after an earthquake, as well as to the future scattering of my ashes across the shortgrass prairie of eastern Colorado. On a broader scale, the amount of disorder, or entropy, has been increasing throughout the universe ever since the Big Bang. If entropy is increasing with the passage of time, ask some devotees of divine creation, how do we explain something as complex as a squirrel? The answer lies in the wider level of chaos in the universe. A squirrel is an island of order whose liveliness is balanced by the increasing disorder in the Sun. Rodent and star are connected by photosynthesis.

Photons streaming from the Sun are the fruits of its decay. These packets of energy reach us in eight minutes and nineteen seconds, Jupiter in 43 minutes and fifteen seconds, and the next nearest star, Proxima Centauri, in a few days short of four years and three months. Around one-third of the thin shaft of light that shines on Earth is reflected back into space, making us visible to the rest of everywhere, and the rest bathes the atmosphere, land and sea. Plants on land and microorganisms in seawater use visible light for photosynthesis using chlorophyll.

The chlorophyll molecule is shaped like a kite, with a flat face that intercepts light and a long tail that keeps it in place inside the cell. Green wavelengths of light are reflected by chlorophyll, which is why plants look green. Chlorophyll is excited by blue and red light and uses the energy conveyed through its structure to blast water molecules apart. This process does two brilliant things. First, it releases the oxygen that we breathe. Second, it generates energized particles called electrons which the plant uses to fuel the capture of carbon dioxide and the assembly of sugars. Sugar molecules are the stuff of life. Plants consume a portion of the sugars that they make by photosynthesis to meet their energy requirements. Some are stored in a sweet form as sucrose, in plants like sugar beet and sugar cane, and others are combined to form bigger and tasteless molecules called polysaccharides that support the body of the plant.

Animals eat plants and turn their substance into animal tissues. This is how the wheel of life turns.

Aquatic microorganisms that also perform the marvel of photosynthesis include algae and certain kinds of bacteria. Marine and freshwater animals depend on these microbes in the same way that the fauna on land relies on plants. The majority of species on land and at sea are tied to sunlight through interactions between the organisms that make sugars and those that eat them. Wildebeests, also called gnus, eat grass, and lions butcher wildebeests. Sun→grass→gnu→lion is a simple chain of custody for the energy originating in the Sun, and is comparable to Sun→algae→krill→baleen whale. Fungi, and many of the kinds of bacteria that do not do photosynthesis, digest the post-mortem remains of plants and animals. But whether the meal is living or dead, its energy came originally from the absorption of sunlight by chlorophyll. Prometheus stole fire from Mount Olympus; chlorophyll takes it from our star.

Nature has also come up with microorganisms that are perfectly content without sunlight. These are called chemotrophs. Chemotrophs make their living by grabbing energy from single atoms of sulphur and iron and simple molecules including ammonia and hydrogen sulphide. They live in great numbers around hot-water chimneys on the deep seafloor called hydrothermal vents, and in less exotic places like animal intestines. Intestinal microbes are interesting because they complicate the transfer of calories from plant to herbivore, and herbivore to carnivore. Wildebeests rely on their gut microbes to digest grasses, and lion intestines cultivate gardens of bacteria that help break down the flesh of wildebeests.

Humans play an outsize role in this terrestrial circus because there are so many of us and because our technological prowess has allowed us to alter the biosphere in ways unavailable to other species. With dominion comes responsibilities, but we have been falling short in our job as custodians. A failure to change our habits will ensure our passage to the thinnest smear in the fossil record. Yet the biosphere will persist. Even if we insist on making conditions inhospitable for all the larger plants and animals, Earth will be cleansed and repopulated by its microbes. We could not extinguish microbiology if we tried. With more than 1 billion years of grace before the Sun starts glowing too bright, there is plenty of time for our home to be remade by the children of the evolutionary future. That would put us in our place.

TWO

GENESIS

How We Arrived

How did we obtain this brief occupation of Earth? Ovid offers two versions of our creation in his *Metamorphoses*. The first scheme has us manufactured from divine seed by a mysterious creator 'who made the world', and the second by Prometheus, the son of a nymph (like Narcissus), who mixed earth with water and moulded us from this clay in the image of the gods: 'And thus the earth, which late before had no shape or hue, / Did take the noble shape of man and was transformed new.'[1] Renditions of this man-from-mud story appear in Sumerian mythology, and feature in the oral history of the Yoruba people of Africa. Dust and clay are the materials used by God in the Bible and Qur'an too. There is a logic and beauty to all of these myths about creation that have us so thoroughly grounded in the substance of this planet. We came from this place, where we were sculpted from the dirt and suffused with life. Until biologists succeed in picking apart the primeval mechanisms that produced the first cells, we have to be content with a cloudy picture of these earliest events. This is like the fog of the Big Bang and the clearness of the physics that follows. What biology can claim is that it has revealed a rich and deeply satisfying account of the birth of animals, including humans, once the first cells were in place. All of life shares this universal nativity.

Our character as a species of ape is a fact of life as certain now as the elliptical orbit of the Earth around the Sun. But as disturbing as our presence in a primatology textbook may be for some theologians, this classification of man does not begin to go far enough. Enlightenment requires deeper inquiry and more imagination than accepting the unsurprising conclusion that animals that look so patently like us are close relatives. To locate our deepest ancestral roots, we begin with a mental exercise in time travel. Joseph Conrad contemplated a similarly challenging odyssey in *Heart of Darkness*: 'Going up that river was like travelling back to the earliest beginnings of the world, when vegetation rioted on earth and the big trees were kings.'[2] But we have to go back much further, to a time when there were no trees at all. If we retreat 100 million years from the twenty-first century, we hear birds chattering at the end of the Cretaceous Period; the next 100-million-year step takes us to the beginning of the Jurassic Period, where crested pterosaurs soar on thermals; another step and we find ourselves walking in hyper-oxygenated forests buzzing with huge insects at the close of the Carboniferous Period; and, one more, to the fish-brimming seas of the Devonian Period, 400 million years ago. We can go back another 100 million years to the freakish fauna of the Cambrian Explosion, but we are still only halfway on the billion-year trek to locate the origins of the earliest distinctively animalish organism.

We have lots of ancestors, lined up in the past all the way back to the mysterious first cells. Around 2 or 3 billion years after the commencement of life, animals sprang from something doubtlessly microscopic. This wriggling forebear was a special organism in the sense that it was the only species whose descendants – all of them – were animals. We have good reason to think that these forebears resembled microbes called choanoflagellates or, more simply, collar flagellates.[3] Collar flagellates live in both salt water and fresh water today. They look a bit like sperm cells, with the addition of a conical collar that surrounds the base of their tail, where it attaches to the body of the cell. The tail is called a flagellum. The collar is made from a ring of microvilli, which are like the little fingers on the surface of cells in our small intestine.

Flagellum and collar work together in the following way. When the flagellum wiggles, it pushes water behind the cell and the cell moves forward (a submarine propeller works in a similar fashion). The water displaced by the flagellum is replaced by water that flows around the cell to blend into the wake behind. On its way, this water is strained through the collar, which serves as a sieve that captures bacteria on its sticky surface. Once immobilized, the bacteria are absorbed into the body of the cell where they are digested. So, the flagellum operates as a propulsion and feeding device. Some types of collar flagellate abandon this free-swimming lifestyle and attach themselves to surfaces with stalks. In these species, the flagellum serves the sole purpose of feeding. Colonial species of collar flagellates develop with multiple cells attached to the same stalk, or swim around as clumps of cells embedded in mucilage.

One of the qualifications for admission into the animal kingdom is that the applicant is multicellular. Zoologists do not admit single-celled designs, and microbes whose cells live in groups are disqualified too. Even if all of the collar flagellates cohabited on stalks or in mucilage globs, they would not get their flagella past the door. There is more to being an animal than a large cell count, including the presence of a blastula or blastocyst stage of development, in which the embryo shapes itself as a fluid-filled ball of cells. All of us did this, five days after the fertilization of one of our mother's eggs. Narcissus was a blastocyst before he possessed more evident beauty, and Joseph Merrick, the Elephant Man, looked equally pretty as an unblemished sphere of 128 cells. We will come back to Mr Merrick in a few pages.

Blastocysts are an important part of the chronicle of our origins. The cells in a blastocyst use special molecules called junction proteins to stick together. (Think miniature rugby scrum, in which the proteins work like the locked arms of the players.) Collar flagellates also produce versions of these proteins, but, significantly, they seem to situate them in their collars, where they filter bacteria and do not use them for joining their cells.[4] If the collar flagellates are related to the ancestors of animals – and everything is pointing in that direction – these proteins may have evolved for netting bacteria and were adapted for gluing cells together later.

Evolution performs this alchemy all the time, modifying single molecules and whole body parts to suit new functions. Feathers, for example, appear to have been used for insulation by the reptilian ancestors of birds. Their recruitment for flight came much later. The difference between this natural re-engineering process and an intentional design is profound. When designing an umbrella, a manufacturer might consider the strength of different materials, how easily waterproofed fabrics can be folded and unfolded, and so on. If natural selection faced the same task, it might start with a bicycle wheel, make the spokes hinge at the hub, and stretch the tire rubber over the whole creation. Elongation of the wheel spindle to produce a shaft and hook handle would be the final step.[5] This is a clumsy way to make an umbrella, but consider the power of evolution which, from such simple beginnings as blotches of pigment on the heads of worms, made eyes for an eagle.

There are three contenders for the title of 'Simplest Animal' and all of them fasten their cells with junction proteins. These are the sponges, comb jellies and a tiny, flattened, worm-like beast called the placozoan. Genetic studies suggest that comb jellies and the placozoan are evolutionary experiments that did not spawn larger branches of the evolutionary tree. It is important to avoid calling these groups 'dead ends' in evolution, because they have done very well for themselves, dodging destruction in successive mass extinctions that wiped out most animal groups. Meanwhile, some ancient sponge, or a close relative to this marine creature, was the ancestor of the rest of the surviving animals, including us. Yes, we are related to bath sponges.[6]

Most sponges live in the sea, feeding on bacteria and other morsels filtered from the water through a system of inlet valves and canals that funnel into an internal chamber. Water that is sucked into this chamber is expelled through a vent called the osculum. The osculum of the giant barrel sponge of the Caribbean is a gaping hole at the top of the animal. The chamber is lined with cells with flagella that sit inside collars. Bacteria are trapped on the surface of the collars and digested within the cells. The resemblance between these cells and collar flagellates was discovered in the 1860s and encouraged biologists to suggest that the flagellates were the forerunners of sponges. But evolution is not as simple as this. The genetics say that the flagellum and collar combination evolved in the common ancestor shared by the flagellates and sponges, and both retained this structure. (Similarly, the presence of enamelled teeth in humans and Komodo dragons does not mean that we evolved from these lizards, or that some of them turned into us; but we do share a common ancestor whose teeth were coated with enamel.)

The reason that the modern biosphere is populated with a mix of single-celled microbes and multicellular beings is uncertain. Although animals and plants with lots of cells have evolved on

Earth, we can imagine life on another planet remaining microbiological throughout its history. Until we can study biology elsewhere in the universe, we will have no way of knowing whether multicellular organisms are bound to arise after the passage of eons of pure microbiology.

Experiments on the behaviour of the collar flagellates may offer some clues about the value of multicellularity. When some of the colonial species are starved by killing the bacteria in the surrounding water with antibiotics, the individual cells in the group separate from one another and wander off. As soon as the solitary cells are reunited with bacteria they form colonies again and even the scent of bacteria is enough to stimulate the same behaviour. Individual cells may be good at searching for food, but a colony can create stronger currents for filter feeding and stay clear of little eddies in the water that might swirl a single cell away. Like a fishing trawler with many crew members, compared to a man with a fishing rod paddling around in a canoe, the colony of flagellates can harvest more prey than a single cell. Multicellularity may have evolved as an effective dining strategy.

Although sponges lack separate organs and have no muscles or nervous system, their anatomy is more elaborate than a mere colony of cells. Like other animals, sponges have multiple cell types that perform distinctive functions. The collar cells are submerged in a jelly produced by specialized secretory cells. Other types of cell are positioned within the jelly and on the outer surface of the sponge. These include cells that make up a simple immune system, contractile cells that close the inlet valves, reproductive cells and cells that secrete the sponge skeleton. Sponge skeletons are made from elastic protein fibres and spines and rays stiffened with silica and calcium carbonate. Silica skeletons of a cold-water sponge called Venus' flower basket are spectacularly intricate structures. Sponges with non-mineralized skeletons of pure protein have been used as bath sponges for centuries. These species supported lucrative businesses in the Mediterranean and Caribbean, until the expansion of this industry ruined the animals and the livelihoods of their human predators.

From an animal that resembled a sponge in its complexity, we trace our evolution through the acquisition of a mouth and an anus in our marine worm ancestry, to fish without jaws, then with jaws, to fish with fins that served as limbs, and on through amphibians and reptiles to something like a tree-shrew, then monkeys and apes. This rich bestiary carried our genes, or genes that would become ours, by cruising the wine dark seas, slithering over glistening bacterial turfs that clothed the rocky shore, exploring the dense jungles that grew in their place and, at last, migrating to the rich savannah where we stood upright in the whispering grass, flared our nostrils to inhale the sweet air of Africa and pondered our next move.

When we scrutinize the genetics of the simplest kinds of animals and their ancestors, we learn something crucial about ourselves. Down at the base of the tree of animal life, deep in the ancestral roots, it becomes impossible to tell the difference between our kin and the forerunners of mushrooms. Animals and fungi merge, forming a stout 1-billion-year-old limb that combines the two.[7] Evidence for this union comes from comparing animal and fungal DNA. This molecular phylogenetic work has been de rigueur for detecting evolutionary relatedness for thirty years. As methods have tightened and data sets have expanded, the case for the affinity between animals and fungi has strengthened. So, beyond the illogic of claiming transcendence over mushrooms that we see on a lawn, we are related to them. We are more similar to fungi than we are to plants or to any of the other major groupings of life.[8]

When a cell is decorated with multiple flagella, biologists refer to the cell as ciliated, and each flagellum is called a cilium. There is, however, no structural difference between flagella and cilia. Ciliated tissues in our bodies include the lining of fallopian tubes, brain ventricles, spinal cord and the respiratory system that, respectively, move eggs, cerebrospinal fluid and mucus. Flagella and cilia are long extensions of the cell membrane that enclose a system of moveable protein rods. The rods slide up and down like pistons, generating the wave motion along the tail. In addition to their use as motors for moving cells through fluids and moving fluids over cells, modified cilia, called primary cilia, are found in almost every type of cell in our bodies. Primary cilia lack a central pair of protein rods and do not wiggle like sperm tails. They work as sensory structures, responding to the mechanical disturbance caused by fluid passing over their surface and enable cells to detect chemicals, light, temperature and gravity.

Defects in flagellar and ciliary function cause a variety of diseases called ciliopathies.[9] Male sterility due to immobilized sperm is the most obvious glitch, but once problems with primary cilia are considered, the range of genetic disorders extends to diseases of the liver, kidney and eye, and to rare syndromes affecting multiple organs. Alström syndrome is a ciliopathy characterized by childhood obesity, vision problems, hearing loss, diabetes and heart failure. It is one of the rarest genetic conditions, with fewer than three hundred cases reported in the medical literature. Marden-Walker syndrome is rarer still, affecting brain development and producing a range of skeletal abnormalities including reduced jaw size, elongated fingers and curvature of the spine. Signs of ciliary involvement are also displayed in cancers affecting the colon, breast, kidney and other organs. These disastrous consequences of ciliary malfunction are caused by the loss of proper responsiveness to various stimuli, impaired signalling between cells and misdirected cell division.

One crucial role played by cilia is to furnish cells with a sense of direction. Determining position might seem unimportant for a single cell, but the mess caused by disorientation is obvious when we consider the careful side-to-side and up-and-down placement of parts as they form in the embryo. The early embryo contains a structure called the node that is lined with a particular kind of motile cilia. Movement of these cilia causes a directional flow of fluid containing signalling molecules. The resulting gradient of molecules stimulates patterns of gene expression that establish the lengthwise left–right axis along which we cannot be folded into equal halves. We are blessed with this *situs solitus* – heart displaced to the left, liver to the right. Errors in bilateral asymmetry – *situs ambiguus* – can affect the function of the heart and many other organs. Interestingly, most people born with *situs reversus*, in which the position of every organ is reversed, live normal lives without suffering any symptoms from their curious developmental history.

Joseph Merrick, the Elephant Man, is thought to have been afflicted with a stupendously rare genetic malfunction called Proteus syndrome.[10] The problem lies in a mutation in a single gene that affects cell proliferation and the process of programmed cell death. Damage to this gene occurs in one cell during foetal development. All of the cells that develop from the mutated cell are affected, while the rest of the cells are unaffected. This results in a mosaicism, in which the individual has a mixture of malformed and normal tissues. The reason for featuring Mr Merrick in this chapter is that Proteus syndrome may be another ciliary defect. Merrick offered the following assessment of his condition:

'Tis true my form is something odd,
But blaming me is blaming God;
Could I create myself anew
I would not fail in pleasing you.
If I could reach from pole to pole
Or grasp the ocean with a span,
I would be measured by the soul;
The mind's the standard of the man.[11]

Long before we occupied the tree of life, single cells that swam the Precambrian seas carried our birthright – genes that would become our genes. This faraway microbial legacy is the first verse of the human saga. It is replayed when sperm cells swarm an egg, and afterwards too, as cells with blunted tails go about their business in every tissue. Whether one's form is something odd, or does not fail to please, it is rooted in the tailed architecture of our cells.

THREE

GUTS

How Our Bodies Work

Following our look at the spongy origins of humanity in the ancient seas, we zip forward in time to the contemporary workings of the body and how this glorious machine walks and runs, sits and sleeps. High on the list of greatest feats of human athleticism must include the record of two hours, one minute and 39 seconds for the marathon set by Kenyan runner Eliud Kipchoge in 2018. The fastest marathon in 1908 took almost three hours, which would be long enough for today's runners to take a tea break at the halfway point and chat about the weather. According to the Greek author Lucian, the first marathon runner, called Philippides, collapsed and died after his exertions in 490 BC. The demise of the Greek herald is not very surprising when we consider that he may have run a total of 240 km (149 mi.) in the days before his 40-km (25-mi.) hotfoot from the Battle of Marathon to deliver news of the Athenian victory.[1] Whatever the body is doing, from running long distances to dozing on the couch, there is a cost to being alive and it all works according to the same rules of chemistry.

Nuclear fusion in the Sun supplies us with food, whether the chain is a short one from potato→human, or a lengthier trip from grass→roast beef→human; or even algae→plankton→small fish→big fish→human. The flow of calories is more complicated for fermented foods and drinks, because yeast is an essential intermediary: grape + fungus→human. With these illustrations of human consumption, it is useful to consider that other microorganisms take the next step by extracting energy from us: human→anthrax bacterium. Infectious microbes have little respect for the claim that *we* occupy the top of any food chain.[2] What we can assert, based on numerous features of our genetics, anatomy and physiology, is a devotion to omnivory. We are generalists and differ from specialists like blue whales and koalas that eat nothing but krill and eucalyptus leaves.

Our nutritional flexibility opens an astonishing range of options at mealtimes. But whether we privilege meat or vegetables, pursue a vegan diet or consume frozen pizzas and orange-coloured snacks sold in metallized plastic bags, the chemistry that releases the calories remains the same. Consider the potato. Capturing energy from a potato is almost as complex as the alchemy performed by the potato plant when it makes potatoes from sunlight, water and carbon dioxide. Potatoes are a good choice for exploring nutrition, because they contain most of what we need to survive. (This does not mean that we would be happy on a diet dominated by potatoes, but we could get by, as Irish peasants were forced to do before things got even worse with the arrival of potato blight in the 1840s.) The potato is made as a hibernation device that enables the wild plant to shrivel its leaves in the autumn, sleep in the winter soil and re-sprout in spring. The tubers contain a good balance of carbohydrates, mostly in the form of starch grains, plus proteins, vitamins C and B6 and plenty of potassium. Fats are absent, which recommends their decoration with butter and sour cream, but even in their unadorned state, mashed potatoes rank as the most filling food available to humanity, according to nutritional research.[3]

We extract energy from mashed potatoes throughout our digestive systems. Beginning in the mouth, human saliva is chock full of enzymes called amylases that break down potato starch into sugars. Enzymes are protein molecules that accelerate chemical reactions that would take years to happen on their own – millions of years in some cases.[4] Other complex carbohydrates are digested in the gut with the assistance of microorganisms which complement the enzymes we produce by ourselves. Bacteria in the gut are particularly good at chopping the bigger compounds released from potatoes into more manageable chemicals. The resulting fuel is swept off in the rich beds of blood vessels running around the walls of the intestine.

Two-thirds of the average distance of 5 m (16 ft) from mouth to anus is taken up by the small intestine. The wall of this part of the digestive system is folded and the surface is covered with tiny projections called villi. Spread over the inside of the gut wall, fields of villi sway with its contractions and the passage of the disintegrating food, resembling the arms of anemones on a coral reef. The surface of each millimetre-long villus is decorated with its own microscopic projections called microvilli. In a healthy gut, the numerous intestinal folds, villi and microvilli make the internal surface up to 120 times larger than a smooth cylinder. Scandinavian researchers have calculated that the digestive area is around 30 sq. m (323 sq. ft), which equals the floor space of a studio apartment.[5] It is all quite miraculous, as long as it works properly. Following a lesson about human digestion in high school biology, one of my friends asked me to ponder the following question: 'What if the intestine is really a giant worm that lives inside us?' He was not the brightest light in the harbour, but I had no comeback.[6]

Worms notwithstanding, most of the nutrients extracted from our food are carried away in the capillaries that loop inside the villi. The capillaries in the gut are part of an immense bed of these minuscule vessels that permeate the entire body. They come within easy reach of all of our 40 trillion cells, furnishing a constant supply of calories, water and oxygen.[7] Capillary beds are connected to our arteries, which carry oxygen-rich blood from the heart, and to our veins, which take oxygen-depleted blood back to the heart. The heart propels blood through 100,000 km of arteries, veins and capillaries by beating 100,000 times a day.[8] Capillaries were discovered by the Italian anatomist Marcello Malpighi, who saw them through his microscope in 1661 in the lungs of a frog. He had experimented on sheep before moving to frogs. Finding that he could not observe the tiniest vessels in the animals while their hearts continued to beat, he found success by removing the lungs and allowing them to begin to dry out and flatten. In the annals of vivisection this was child's play. Greater savagery came at the hands of William Harvey, the English physician, who understood circulation by watching it in dogs and deer strapped to tables, their necks and chests opened to the sky.

After eating a serving of mashed potatoes, our red blood cells load up on glucose released from the potato starch and are pushed on through the bloodstream. Hungry cells throughout the body absorb these sugars from the closest capillaries. Oxygen is needed to release energy from the food and this enters the blood vessels from the alveoli of the lungs. (Malpighi discovered the alveoli too.) Once the sugar molecules are inside the cell, they are broken into smaller parts and energy is harvested from the fragments by pulling electrons from their component atoms.[9] Sugar metabolism occurs in distinct steps controlled by specific enzymes. Many of the enzymes are organized inside separate structures within cells called mitochondria. Drawings of cells illustrate mitochondria as pill-shaped blobs with wrinkled interior membranes. Most of the energy in our food is captured through the process of oxidation in the mitochondria.

Life is a slow burn. This metaphor has more than poetic meaning. The body consumes oxygen like a bonfire and leaves nothing but water and puffs of carbon dioxide behind. The difference lies in the way that energy is released during the burn. When logs are blazing, electrons are stripped from molecules in the crackling wood, and water vapour and carbon dioxide are carried away as the flames leap into the air. Most of the energy in the wood is released as infrared radiation or heat, with the visible light of the flames representing a secondary form of energy emission. Oxygen does the same job in the bonfire as it does in the cell – namely, grabbing electrons from the materials being oxidized. The controlled way in which sugars are oxidized in cells contrasts with the uncontrolled fury of a fire. Much of this is due to the stepwise breakdown of sugars in living things and the way that the reactions are separated into different compartments inside cells. This tightly controlled process allows the cell to harvest a good deal of the energy in the form of chemicals that it uses as a portable fuel. Nevertheless, the tiny mitochondrial furnaces lose energy by heating to 50 degrees Celsius (122 degrees Fahrenheit) as they burn through the sugar.[10]

Getting sugars into cells is a crafty business. Cells are enveloped by lipid membranes which serve as water-resistant barriers that separate them from their surroundings. Lipids are oily molecules that do not dissolve in water. Cells in the liver are surrounded by other liver cells; blood cells are surrounded by the fluid plasma of the bloodstream; single-celled organisms like amoebae are surrounded by pond water. Chemicals move in and out of cells, but sugars and other things that dissolve in water cannot move freely through the membrane. Chemicals are free to diffuse through water in a pond, but they are not free to enter or leave a cell. This enables the amoeba to keep a tight rein on every facet of its make-up. It exists as an island of order surrounded by the disorder of its pond.

Cells are comparable to very tidy houses, in which the walls define an ordered living space and doors and windows control what comes in and what goes out. Cells regulate their contents with proteins situated in their membranes that act as gateways for the passage of substances dissolved in water. Single atoms as well as larger molecules cross through these proteins. Glucose is imported via proteins that flex open to offer a pocket of the perfect size to usher sugar molecules across the membrane. Other transport proteins create electrical voltages across membranes as they push charged ions, including sodium (Na^+) and potassium (K^+), from one side to the other. We are familiar with voltages produced by batteries. These are established by the flow of charged electrons and ions between different metals. This can be demonstrated by plugging copper and zinc wires into a potato and using the resulting current to run a digital clock. The same principles apply to individual cells, which operate as miniature batteries. Voltages across membranes are the essence of life because they power the uptake of glucose and other substances. Mitochondria also work as batteries inside the cells that house them, converting voltages across their wrinkled inner membranes into the energy used for chemical reactions. Chloroplasts are batteries too, although they are charged by sunlight, like solar panels. Life draws on the battery power of cells.

Nerve cells, or neurons, are electrified by the movement of sodium and potassium ions through proteins in their membranes. The protein channels open and close along the length of the nerve cells, causing changes in the membrane voltage. These electrical impulses travel along nerve fibres and are propagated from one cell to the next via connections called synapses. Each neuron forms multiple synapses, which configure the nervous system as a network, labyrinth upon labyrinth rather than bundles of straight pipelines.

The neocortex is the outermost part of the mammal brain forming the familiar corrugated surface. Ours contains 16 billion neurons connected by 100 trillion synapses. With this circuitry we laugh and cry, fall in love and into despair, write epic poems and pitiful tweets. It is the source of our artistic and scientific breakthroughs, as well as the gnawing narcissism of the species. Incidentally, sperm whale brains are six times bigger than human brains and the neocortex of the long-finned pilot whale packs twice as many neurons as you.[11] What songs of love and despair do cetaceans sing in the wine dark sea? In itself, the presence of a neocortex is a poor guide to smarts. Lacking this evolutionary add-on, African grey parrots and octopuses solve all manner of complex problems. According to some animal psychologists, a famous parrot, called Alex, had a vocabulary of more than one hundred words, could describe the size and colour of objects and carry out simple computations.[12] The boredom expressed by octopuses in captivity is a sure sign of their intelligence, and there are stories of the animals squirting aquarium workers for entertainment, others juggling hermit crabs to pass the time, and the most determined creatures executing elaborate escape plans.

One thing that nervous systems are very good at is making us mobile, coordinating our conscious and subconscious movements. This was particularly important before the invention of agriculture, when we had to catch things from the wild. Some anthropologists think that we seized antelope and other meaty animals by outrunning them – not by sprinting, but by marathoning them into exhaustion. This is called persistence hunting. Wolves, wild dogs and hyenas do the same thing. Humans can be very good at this, outlasting prey in the heat of the day by regulating body temperature through sweating and continuing to run without eating or drinking.[13] Early humans used weapons too and probably trapped animals in pits.

Another school of anthropological thought promotes the role of scavenging in human evolution, with our ancestors showing up at a feast after a kill by a more proficient predator. Given that a carnivore like a sabre-toothed tiger or a lion would treat us as another meal, we may have tracked the big cats, kept our distance and eaten whatever they left behind. This picture is at odds with the notion of man as the supreme hunter, but it is a good bet that we pulled the stringy bits from big herbivores after the lions had torn open the body cavity and gorged themselves on the glistening offal. Early arrival was the key. Rotting carcasses would never have been appealing because the microorganisms doing the rotting produce lethal toxins. Alligators and vultures that specialize in feeding on decomposing flesh are equipped with super-powered gastric juices and have mixes of gut microbes that provide resistance to poisoning and infection from the putrefying meat.[14] Humans pursued the more common evolutionary approach of avoidance, which may explain why we are so sensitive to the smell of decomposing meat.

Protein digestion begins in the acid bath of the stomach and continues in the small intestine. Enzymes release amino acids from the proteins and these are processed in the liver before they are oxidized in the mitochondria like sugars. Fats are digested in the small intestine, releasing fatty acids, and these are also burned in the mitochondria. But glucose released by starch digestion is the perfect fuel for the body. We have multiple copies of a gene that codes for the amylase enzyme that breaks down the elongated starch molecules, releasing individual glucose sugars like pearls from a strand. This allows us to produce more of the enzyme in saliva than other species of apes equipped with only one or two copies of the gene. Cooking tubers on fires changes the conformation of their starch grains and allows the amylase enzyme to release sugars more readily. The combination of these genetic modifications with the technology of fire may have furnished us with the energy needed to develop our big brains. We did not abandon meat, but some anthropologists believe that cooking starchy vegetables was the critical change in behaviour that facilitated brain expansion.[15]

A brain draws around 20 watts of power, which is the same as a compact fluorescent bulb that produces as much light as the heated tungsten filament in an old 100-watt bulb. For its size, the brain is a big drain on energy, with the rest of the body requiring another 80 watts to keep going at a sedate pace. A good deal of this energy is released as heat, which is one of the reasons that crowded rooms become so uncomfortable. The necessary vitality comes from consuming at least 2,000 food calories per day, which could be achieved by eating seven large potatoes (2.6 kg; 6 lb) or one plateful of steak (just under a kilogram; 2 lb). In common with other primates, we thrive on less than half of the energy compared with less efficient mammals of the same size.[16] If we consider that Titian and Francis Bacon painted their masterpieces by consuming no more power than a light bulb, we have a solid scientific reason to celebrate the fruits of civilization. But, when we ponder how much additional energy we consume as we go about our business in the twenty-first century, we lose some of our objective charm. The average American consumes 12,000 kilowatt hours of electricity in a year, which is associated with the annual release of 16 tons of carbon dioxide into the atmosphere.[17] This level of individual energy use is matched by two-and-a-half British residents, fifty Mauritanians and more than 340 people in the Central African Republic.

In addition to supplying the brain and other body parts with fuel, there are lots of metabolic expenses associated with keeping noxious microbes at bay. Throughout our journey from uterus to grave, microorganisms are intent on turning off our lights. Because nature abhors a vacuum – *horror vacui* – the aggravation of every larger organism by a galaxy of bespoke infectious microbes is an unavoidable nuisance.[18] Not that it changes anything in practical terms for the patient ravaged by a brain infection, but there is no malevolence at work here. Like us, bacteria and the more numerous infectious particles of viruses convey their genes through time.[19] If they did not do so they would not be here, which is the simplest telling of natural selection. Survival is reserved for effective genes in effective cells and effective bodies. (The word 'adequate' could substitute for 'effective' here, because adequacy to the task of survival is all it takes.) The human response to the superfluity of invisible monsters is the miracle of the immune system. (Miracle in the sense of stunning, rather than the product of sorcery.) Our tissues are policed by cells of the immune system, including white blood cells that sniff out germs, gulp them through their reactive surfaces and destroy them once inside. Some immune cells specialize in identifying unwanted microbes and signalling to other cells that operate as executioners.

The body can be its own worst enemy when we spawn cancer cells that abandon the graces of cooperation and multiply at the expense of surrounding healthy tissues. Their formation is a troublesome consequence of the error-prone copying of DNA that refreshes our tissues. Experiments on mutant mice lacking components of their immune systems suggest that cancer cells are spawned every day.[20] Cancer is an ingredient of every life and functioning immune systems scrub these misbehaving cells from the body.

When we consider the cacophony of chemical conversations in the body, from human cell to human cell, between the microbes of the microbiome, and between these bacteria and our cells, it is clear that we are vibrant, travelling ecosystems, apes that bear communities of microbes on us and inside us, veritable coral reefs of biological diversity, without the splashy colours. We used to carry more in the way of skin lice, intestinal worms and other parasitic animals, but their loss is preferable, on balance, to life with a relentless crawling sensation. As many as 30,000 lice can colonize a single person. In a memorable account from the twelfth century, legions of the parasites 'boiled over like water in a simmering cauldron' from the infested clothing of Thomas à Beckett, Archbishop of Canterbury, as his body cooled following his assassination, 'and the onlookers burst into alternate weeping and laughter'.[21]

And so, what should we judge from our reflections on the full comprehension of the bodily machine? *Homo* somewhat sapient, but still this mineral frame, strung with bands of protein and smoothed with blobs of fat, electrically wired, aerated by bellows in the chest, nourished and drained via an elaborate plumbing system, appended with organ meats and wrapped in an elasticated hide.[22] 'What a piece of work . . . the paragon of animals,' as Hamlet put it (Act II, Scene 2).

FOUR

GENES

How We Are Programmed

Genes instruct the assembly of organisms that convey copies of genes to the next generation. We are temporary vessels for genes, situated in family trees that assume the shape of a river delta, with DNA streams draining down from ancestors to descendants. These currents merge whenever a sperm cell fertilizes an egg, rousing the delta with an opportunity to gush on; without children, DNA swirls in its channel before silting up. People with religious faith believe that they have been awarded a purpose in life that transcends the pure genetics of existence; those that put these longings for an afterlife aside must be content with the poetry of flowing DNA. Neither viewpoint assures contentment, but the Sun also ariseth, as it says in Ecclesiastes, the cat also meoweth to be let out and the fate of tomorrow is never settled today.

Human genes are distributed among 23 pairs of chromosomes packaged in the cell compartment called the nucleus, plus copies of a tiny accessory chromosome carried by each mitochondrion. The complete set of chromosomes makes up the genome. Chromosomes are made from pairs of DNA strands. These are held together by rungs to produce ladders that twist into the shape of the iconic double helix. The DNA is wound around special proteins called histones, and further compacted to fit inside the nucleus. If the longest chromosome was unravelled, its helices would extend for 85 mm (3.3 in.), and the DNA in all 46 chromosomes would cover 2 m (6.6 ft). The nucleus is only a few millionths of a metre in width, but the DNA strands are very thin and are packed very tightly. This is evident from the fact that a model of the human genome in which the DNA helix was enlarged to the thickness of a pencil would stretch for 8,000 km (5,000 mi.).[1]

Genomes contain all of the information for manufacturing their organisms, but can only get to work inside cells. A genome cannot make a cell from scratch, let alone a whole multicelled organism. This interdependency of cells and their genomes is a vital fact of life. Long before the existence of genes was imagined, seventeenth-century naturalists became convinced that *omne vivum ex ovo* – or, all life (comes) from an egg.[2] Two centuries later, the egg maxim was rolled into the cell theory as *omnis cellula e cellula* – every cell from a cell. At least one cell balked this rule: having no cellular parent, the first cell on Earth arose from chemistry rather than biology. Since this epochal event, genes have been transferred from cell to cell, from one generation of organisms to the next, for billions of years, via unbroken lines of custody from microbes to the multicellular animals, plants, seaweeds and fungi that fill the tree of life. All of these organisms, from the smallest to the largest – from microbes called mycoplasmas to blue whales or gigantic mushroom colonies – are programmed by their genomes.[3] The differences between whales and worms, and worms and everything else, lie in their genes. There is no other source of information on which to formulate an animal.

The overwhelming complexity of the processes of biology urges us to describe life as a miracle, but we have to put this notion aside to understand ourselves. The sight of a newborn baby provokes wild swings of emotion, but there is nothing that is extraordinary about this for anyone beyond the family of the infant.[4] When we consider the brilliance of the latest mobile phone or commercial aeroplane we have little idea how they work, but we are certain that they were assembled on production lines by skilled tech workers and engineers. There is an innate resistance to applying the same reasoning to a baby, although we are confident that it was made from a fertilized egg in the mother.

Babies are fabricated according to instructions in their genomes, with the support of the genomes of their mothers that prepare the womb, make one side of the placenta and feed the foetus.[5] The format of the directions for making mobile phones and aircraft is very different from those composed by nature. The last steps in the installation of the propeller of a Cessna 172 – the world's best-selling plane – read 'tighten propeller attaching bolts' and 'install stainless steel safety wire'. Notice that the engineer is told what to do and the instructions assume that she, or he, knows what tools to use. If biology attempted to build a Cessna 172, the DNA would need to do a lot more than this. Cessna DNA would detail the manufacture of every component of the plane and also make sure everything was attached in the right place. Our genome specifies the composition of tens of thousands of different kinds of molecule and instructs them where to go and what to do. Millions of genes would be needed if each step in the assembly and operating processes for a human cell required its own instruction. But tasks are streamlined in the genome and just 20,000 genes get the job done.

DNA is able to store so much information because it commands the formation of very accomplished robots that get on with things without constant supervision. These robots are enzymes that accelerate chemical reactions and each enzyme performs its task according to a programme that is built into its structure. Returning to the Cessna analogy, a tool that worked like an enzyme would attach itself to a bolt automatically and immediately tighten it on to the propeller. Rather than specifying 'tighten propeller attaching bolts', the manual would read, 'Make Tool Alpha'.

The reason that enzymes are so good at their jobs is that they have been tested throughout the history of biology. Some of the oldest enzymes carry out housekeeping tasks in all organisms, like releasing energy from sugars. With minor modifications, these enzymes have survived continuous appraisal for billions of years. If a version of one of these tools is produced in an organism and works so poorly that this individual does not reproduce, the instruction for making that enzyme – namely a gene – is not transferred to the next generation. It perishes with the failed experiment that carried it. If another version works better than the original – better enough to increase the likelihood that the bearer of the gene will leave offspring – it will prosper and may even replace the original gene in a few generations. This testing and modification of genes is the essence of evolution.

Enzymes are proteins and proteins are assembled as strings of amino acids. There are twenty different amino acids and DNA specifies the order in which they are attached. Genes are written in the familiar letters of the DNA code, the As, Ts, Gs and Cs, which are molecules that form the rungs on the DNA ladder. Amino acids are spelled out in three-letter words written in this four-letter alphabet: AAG codes for an amino acid called lysine, for example, and GCA for alanine, which means that a DNA sequence AAGGCAAAG specifies lysine-alanine-lysine. A human protein of average size contains more than four hundred amino acids. Once we recognize that sequences of the letters in DNA determine the structure of enzymes, it is clear why mutations that change these sequences affect the way that enzymes work.

In addition to itemizing enzymes, DNA sequences define proteins that form mechanical frameworks inside cells, receptor proteins that receive and transmit chemical messages, and a variety of proteins that regulate the expression of genes. Beyond their bounty of proteins, cells have lipid molecules in membranes, lots of the sugary compounds called polysaccharides, as well as DNA and other nucleic acids. All of the non-protein components have to be listed in the genome too, but this is done indirectly, by specifying the enzymes that make them. Cholesterol is a type of fat molecule that is assembled by the action of multiple enzymes encoded by separate genes.

The human genome is written with 3 billion letters and is very cluttered. Our 20,000 genes occupy less than 2 per cent of this script. All of them could fit on a single chromosome with lots of room to spare, but evolution is indifferent to tidiness. Because the human genome was crafted from the earlier genomes of our ancestors, it carries the faultless as well as the corrupted genes from this history, an inheritance that runs all the way back through fish to bacteria. This is why the genes that make our proteins are scattered across the chromosomes and planted in deep thickets of meaningless text. The messy majority of human DNA is written as strings of letters that code for no amino acids at all, or script that codes for chains of amino acids that make proteins that do not work. Some of this DNA performs useful tasks without making any proteins, but most of it seems to be rubbish, or junk DNA. As time passes and natural selection pans for gemstones – the next best enzymes – the accumulation of these mine tailings is unavoidable.

The human genome is unexceptional in its informational capacity. We have three times as many genes as yeast, the same as roundworms and chickens, but fewer than many plants. The largest known genome belongs to a Japanese lily called the canopy plant.[6]

The single white flower of this exotic plant sits on top of a whorl of bright green leaves. It has fifty times more DNA in each of its cells than we do. Even with the rapid sequencing methods available today, it is unpractical to read all of the letters in such an ungodly dollop of DNA, so the number of lily genes is unknown. Wheat, which has a smaller genome, has an estimated 95,000 genes. The massive DNA endowment of wheat arose from the amalgamation of genomes from three wild grass ancestors, and all of them had a bigger genome than us to begin with. Among animals with big genomes, the marbled lungfish from Africa has almost as much DNA as wheat.

Before the completion of the human genome project, biologists felt quite confident that we might have as many as 100,000 genes. Participants in the sequencing work were flabbergasted as the likely number fell to 30,000 by the time of publication. The following response from a prominent French geneticist in 2001 in the American journal *Science* expresses their surprise:

> That a mere one-third increase in gene numbers [should read one-half] could be enough to progress from a rather unsophisticated nematode [with about 20,000 genes] to humans . . . is certainly quite provocative and will undoubtedly trigger scientific, philosophical, ethical, and religious questions throughout the beginning of this new century.[7]

As subsequent pruning took the number down to 20,000 to match the nematode worm, most biologists had already begun to rethink things. But more than a decade later, a desperate band of geneticists – absent bandoliers – cast reason to the winds and proposed that a treasure trove of information lay hidden in junk

DNA. They became entranced by the possibility that the ostensibly silent majority of the human code contained an encyclopaedia of instructions that did not work through protein synthesis. Hundreds of millions of dollars in the United States were invested in sifting through the junk for little profit.

When a gene, written in DNA, instructs protein formation, it is read, or transcribed, into a different kind of nucleic acid called RNA. This RNA acts as an intermediary between the gene and the cell hardware that manufactures proteins. It has been known for a long time that RNA can do a number of things besides its role in making proteins. The idea behind the belief in secret messages in junk DNA was that the messy sequence generated a lot more of this RNA activity and that this was the source of all the great stuff that made humans so fantastic.

And then came the 'Onion Test'.[8] Onions have five times as much DNA as humans. Onions are among the wonders of creation, particularly as they sizzle in olive oil, but does it really take a lot more DNA to make this vegetable than a human being? Without succumbing to narcissism, it does seem more reasonable to conclude that onions, like us, have lots of junk DNA. Critics of this viewpoint include religious creationists, who are committed to the idea of human specialness. They are the ultimate egotists. Not content with loving themselves in private, they believe that humans are specially designed and favoured above every other resident of the biosphere. They are content with the idea that onions maintain a heap of genetic refuse, but have difficulty accepting the notion of human junk being junk.

The relationship between the complexity of an organism and the size of its genome is very weak. By complexity, we tend to think about the size of an organism, its anatomy and what it does for a living. There is certainly more to being a human than a nematode worm, but not much more, perhaps, to being a cell in either organism. This is an important point, so much so that it deserves to be recast for emphasis: humans have more moving parts than worms, but the individual cells that make them are equally complicated.

Think about building a Cessna versus an Airbus A380, the largest passenger airliner. There are, obviously, more steps in the construction process for the Airbus, but many of the tasks are repetitive. In both cases, some components are attached with bolts using a torque wrench. With the genome approach, the instructions are streamlined because most genes instruct the formation of tools and those tools are smart enough to get on with their jobs automatically. Worms and humans need equally large repertoires of genes to run their metabolism, shape their anatomy, coordinate their movements, allow them to digest food, furnish them with immune defences and so on. The differences between us appear to reside in a relatively small subset of genes.

Brain size is an obvious variance between worm and man. The nematode brain consists of a ring of nerve cells surrounding the front end of its digestive tube. It is an abacus next to the supercomputer in our skulls. The human brain is a big organ and its presence in an animal with an upright gait, varying levels of athleticism, opposable thumbs, decent vision and a marked tendency towards violence has allowed us to swan around as if we owned the place. The genetic foundation for these uniquely human attributes appears to be slender. We have identified a few genes that may account for our large brains and these are absent from worms and also from our closest relatives, the other great apes. These have been formed by the duplication of existing genes, followed by modification to assume new roles. Gene duplication is a prime mover in evolution because it is the source of additional informational capacity that can be turned to new tasks. Two of these novel genes that are related to brain development affect the growth and development of nerve cells. When they are inserted into the mouse genome, the rodent brains acquire a thicker set of connections and their surface becomes more folded. Confined to their cages, gripping the wire with their little pink claws, twitching their whiskers and blinking their black marble eyes, I wonder if the outlook of the captive rodents becomes darker or brighter as their brains wrinkle?

The human genome has been reworked since we left the Rift Valley more than a thousand centuries ago. All of us carry mutations in our DNA in which letters in the sequence are swapped – a C replaced with a T or an A, for example. Most of these modifications cause no harm, but some serious genetic conditions have this basis too. Tay-Sachs disease and sickle cell anaemia are examples of illnesses caused by a change in a single letter in a gene sequence. Every human being is stocked with a unique set of these fine-grained changes to the genome. Any pair of humans may differ at 4 to 5 million spots in their genomes, which sounds like a lot but means that their DNA varies by less than 1 per cent.[9] Even identical twins differ at this level of detail because these mutations can occur after their common embryo has split in two. People with family roots in Europe or Asia show fewer of these refinements in their DNA than Africans.[10] This supports the well-established African origin of modern humans: the oldest populations on Earth harbour the greatest genetic diversity.

These modifications reveal other features of human biology. First, human genetic diversity is very low compared with most animal species; Norwegians and Nigerians share almost all of their DNA. Second, few of the differences between individuals bear any relation to the racial categories that appear on census forms. The genetic differences that account for variations in skin colour and other physical features between native Scandinavians and Africans are founded on very few genes. Our geographical distinctions are very small in comparison with the larger variations that pepper everyone's genome.

Carl Linnaeus gave us our Latin name in the eighteenth century and split *Homo sapiens* into a quartet of geographical varieties according to skin colour and perceived differences in behaviour. Founded on this *racial* classification, later scientists created more overtly *racist* taxonomies that situated whitewashed Europeans as the pinnacle of the species and presumed that other races had formed by degeneration. Genetic research has squashed these beliefs, but their persistence among large swathes of humanity is one more manifestation of our narcissism. Not content with placing the rest of nature on a sliding scale from delicious to detestable, hierarchies are imagined within our own species. These convictions come very cheaply. Racism can serve as a convenient refuge for people who have failed to identify any trace of personal worth.[11]

FIVE

GESTATION

How We Are Born

Two hundred and fifty babies are born every minute. The frequency of the process argues against calling it a miracle, but the spellbinding sight of a newborn can excuse this sloppy expression and mothers deserve veneration for their pains. After nine months of concealed assembly, there it is, rubbery and glistening, leaving us awestruck by the workings of nature. And awesome they are. If the arrangement of your adult organs allows you to inhale and exhale, digest food and urinate, it is certain that the chemical reactions that criss-crossed your unfolding embryo performed greater magic than the wildest designs of a modern bioengineer.

Like other sexual animals, we begin as a merger of two types of cell. Lured by pulses of ovular perfume, hundreds of sperm cells thrash around the egg. One from the swarm nudges between the obstructing follicle cells, spits a drop of enzymes from its head to dissolve a path through the coat below and sticks to the egg membrane. Fertilization proceeds with the passage of the nucleus from the sperm into the egg. Within 24 hours, the fertilized egg divides in two. Subsequent divisions create a ball of cells and, when about thirty cells have formed, the congregation organizes itself into a fluid-filled sphere called the blastocyst.[1] The structure of a blastocyst is as simple as a colony of pond algae.[2]

Greater complexity begins to emerge when an inner mass of cells forms at one end of the blastocyst and the embryo is planted in the wall of the uterus. The blastocyst becomes a gastrula when the layout of the body is established. This part of the engineering operation begins with the formation of a groove on the side of the embryo that will become the back of the animal. The groove is part of an emerging structure called the primitive streak. This serves as an early guidance system to ensure that the head ends up at the opposite end of the animal from its anus – which has its advantages. The left and right side of the body are positioned on either side of the streak.[3]

Gastrula formation is also associated with the formation of three distinct tissue layers. The outermost layer, called the ectoderm, forms the skin and the nervous system; the middle layer, or mesoderm, is the source of muscle and bone tissues, and the gut and lungs come from the inner endoderm. A rod called the notochord forms inside the gastrula as the tissue layers are laid out. (Later in development, this flexible stick is absorbed into the backbone with the formation of the bony vertebrae.) A flattened plate of cells develops at one end of the primitive streak, which elongates and folds in on itself to create a tube that houses the nerve cord, which will become the spinal cord, and the swellings that become the brain. The embryo is no bigger than a sesame seed at this time, long before the identity of the animal becomes obvious.

Although the anatomy of vertebrates – animals with backbones – is more complicated than the structure of worms and insects, there are many similarities in the construction process. Earthworms have obvious segments, evident as rings on the outside and repetitive body parts on the inside. Modifications to a standardized segment plan customize the body components from one segment to the next. In the worm nervous system, for example, a pair of brain swellings at the front end arise from less conspicuous bulges that repeat along the nerve cord in the other segments. The same goes for insects. The bodies of the larva and adult honey bee are organized as a series of segments visible as the rings on the outer skeleton of the animal. Mouthparts and antennae are attached to the ones at the front end of the adult, with three pairs of legs further back. The segmented nature of vertebrates is less obvious, but is apparent in the stacking of vertebra along the spine. Each vertebra corresponds to a body segment, referred to as a somite in the embryo. There is a common body plan at work here, accommodating hundreds of vertebrae and ribs in snakes, and reduced to 33 vertebrae and a dozen ribs in us.

Every cell in an embryo contains the whole genome of the organism. The reason that brain cells work differently from lung cells is that specialized subsets of genes are active in the two types of cell. As the embryo grows, *Hox* genes turn developmental pathways on or off according to the functions of each body segment. *Hox* genes are organized along the length of chromosomes in the order that they are expressed, beginning with the genes that affect the formation of the head, followed by other *Hox* genes that control development towards the tail of the embryo. This arrangement of genes aids their expression in the correct segment-by-segment order. Errors are disastrous. Mutations in developmental genes in fruit flies swap their antennae with stumpy legs, warp their translucent wings and shrink their eyes to dots. The consequences of these mutations in vertebrate animals include abnormal limb development and disordered organs, facial deformities, cancer and hearing loss.

The study of developmental abnormalities is called teratology and the heartbreaking specimens of this science are displayed in jars in anatomy museums. Embryologists are enthusiastic about interfering with the embryos of chickens and other animals, but our knowledge of human teratology relies on the analysis of natural programming errors like *Hox* gene mutations. A fourteen-day rule allows research on human embryos before they form the primitive streak and the left–right and head–tail organization begins to emerge. Eggs fertilized in the lab and cultured in Petri dishes will develop normally for one week, producing a blastocyst that can take root in the uterus if it is transferred into a prospective mother. New culture methods have succeeded in allowing rather misshapen embryos to grow for thirteen days after fertilization.[4] The promise of these techniques has led to calls to revise the current legislation, but there is significant opposition from ethicists.[5]

After the formation of the neural tube, all vertebrate animals pass through a developmental stage in which they look surprisingly similar. Fish, amphibians, reptiles, birds and mammals appear quite fishy – like fatty seahorses.[6] Decades before the publication of Darwin's evolutionary theory, study of the fossil record had shown that fish had evolved before other vertebrates. This encouraged ideas about a clockwork procession of beasts onto the land, the rise of the dinosaurs, birds and, crucially, Victorian gentlemen. Belief in our creation as the high point of evolutionary progress also encouraged fears about a reversal in fortunes – the possibility of bestial relapse. Robert Louis Stevenson fed this disquiet in his novel *The Strange Case of Dr Jekyll and Mr Hyde*, published in 1886:

> that insurgent . . . lay caged in his flesh, where he heard it mutter and felt it struggle to be born; and at every hour of weakness, and in the confidence of slumber, prevailed against him, and deposed him out of life.[7]

Early embryologists thought that they could see evidence of evolutionary progress in the embryos of different species, and this led to the claim that all vertebrates passed through a fish stage of development. The resemblance is real, but the modern reading of embryos reveals patterns of common ancestry: we all evolved from the same ancient wormy ancestor. The facts show, then, that salmon pass through a stage of development in which they look like the embryos of cheetahs, eagles resemble frogs and so on.

Among the most prominent shared features of embryos in this phase of manufacture are the folds that develop below the head. These are called pharyngeal pouches. Slits develop from the folds between the pouches in fish and poke through to create gills. Gill slits do not form in land animals and, instead, the pouches become very important in the development of the body segments. Part of the middle ear and eardrum form in the uppermost of the pouches in mammals; the thymus gland, which makes the protective T cells of the immune system, develops in the third and fourth pouches. At the other end of the embryo, the growth of a tail bud adds to the features that make it difficult to tell embryonic lizards from zebra. The eyes, the heart and other organs, the gut and four limb buds are carved out at this time too. The heart begins to beat and the embryo begins to assume the form of the animal that will be born months later.

The way that the identity of different animals unfolds from the common plan of the embryo is overwhelmingly beautiful. Bat fingers elongate and stretch the wing skin between them; elephant noses and upper lips are joined into little trunks; giraffe embryos grow long necks and dainty hooves. Wave upon wave of gene expression results in the precision sculpture of each mammal in its womb. These defining modifications continue after birth and the skeletons of adult mammals reveal how many of the seemingly profound differences between species arise from the shortening and lengthening of a common set of bones.[8] Humans develop in less than half the time it takes to make an elephant. Small rodents are the swiftest of the assembly jobs among the mammals with placentas, with shrews escaping the womb in two weeks. Marsupials are born in as little as twelve days, but they are helpless pinkies, tiny as bees, and spend the next few weeks in their mother's pouch.

Whale and dolphin development takes the prize for breathtaking embryological transformation, with the absorption of the hindlimb buds and flattening of the forelimbs into fins. When everything goes to plan, a sperm whale mother gives birth to a 1-ton, 4-m-long (13-ft) calf, which is nudged to the surface by other members of the pod to take its first breath of briny air. Sperm whale gestation lasts for fourteen to sixteen months, and the calf is nursed for two years. Hermann Melville described calves newborn and *in utero* in *Moby-Dick*:

> One of these little infants, that from certain queer tokens seemed hardly a day old, might have measured some fourteen feet in length, and some six feet in girth. He was a little frisky; though as yet his body seemed scarce yet recovered from that irksome position it had so lately occupied in the maternal reticule; where, tail to head, and all ready for the final spring, the unborn whale lies bent like a Tartar's bow. The delicate side-fins, and the palms of his flukes, still freshly retained the plaited crumpled appearance of a baby's ears newly arrived from foreign parts.[9]

All mammals arrive from foreign parts and have no memory of those watery months. I was born five years before abortion was legalized in the United Kingdom in 1967, and have good reason to believe that my biological mother would have opted for a termination had this been available. Vacuum aspiration would have sucked me into early oblivion and my adoptive parents would have been offered a different child. It is unsettling to consider this, but some objectivity changes the picture. The embryo that might have been terminated was not me; it was the seahorsey stage of a mammal that became me. If the abortion procedure had been carried out later in the pregnancy, the foetus would have looked a lot more like a newborn baby, but that would not have been me either; it would have been the foetal mammal that became me. A mammal is anticipated in the womb, but the individual – the person – does little to express itself until, like the whale, it takes its first breath.

Edmund Spenser offers a lyrical expression for the limits of this sort of counterfactual thinking in *The Faerie Queene*:

> As when a ship, that flyes fayre under sayle,
> And hidden rocke escaped hath unwares,
> That lay in wait her wrack for to bewaile,
> The Marriner yet halfe amazed stares
> At peril past, and yet in doubt ne dares
> To joye at his foolhappie oversight . . .[10]

What was one of your hidden rocks? How about the time the side mirror of an accelerating bus whizzed by at head height the moment before you stepped off the curb? How lucky that a few seconds earlier you were distracted by a yellow moth fluttering behind the window of the sandwich shop and slowed your pace. Did the insect save your life, or should you thank the cleaner who did not bother to open the window? Life proceeds as a continuous stream of near misses and then it stops. The possibility of an intentional abortion is an example of an early crossroad. But what if you had died during birth, your embryo had aborted spontaneously, your fertilized egg had failed to implant, or if your parents had not had sex on that seemingly consequential day? And, with all of the uncertainties of life, one of these early interruptions might even prove the best of all possible outcomes, saving a future family from losing a beloved parent when the side mirror of a bus did make contact.[11]

Distressed by the unpleasant rules for mankind after the Fall, Adam questioned God's motivation for his creation in Milton's *Paradise Lost*:

> Did I request thee, Maker, from my clay
> To mould me Man, did I solicit thee
> From darkness to promote me, or here place
> In this delicious garden? (Book X, 743–6)

Mary Shelley used Adam's plea to great effect as the opening epigraph of *Frankenstein* (1818). Swept away by the furnace of his egotism, Victor Frankenstein had expected his monster to be grateful. At its best, life is a very surprising gift; at worst, an unwelcome burden. Considering the fortunes of my life, it would be difficult to regret being born. On the other hand, if 'my' fertilized egg had vanished I would not be here to express remorse and I would never have been known to the handful of people who lay some value in my respirations. The estimated number of induced abortions each year is 60 million. If abortion was eliminated, the annual growth of the global population would shift from a little above 1 per cent to 2 per cent. We can imagine a possible citizenry that would include individuals who did not make it beyond the womb and it would look very much like us – more of much the same, with no change in the proportions of poets and fools.

For some opponents of abortion, the thought of all those prospective unborn babies is unbearable. Abortion looms large in their imagination and questions about abortion access dominate their choice of political candidates. Their interpretation of religious ethics may go further, forbidding any form of contraception. These views are sermonized as a love of life, of any and every human life. Issues relevant to the wishes of a mother, a threat to her health caused by the pregnancy, or the detection of severe foetal abnormalities, are insignificant compared with the value of carrying every baby to term. This is certainly the position of the Vatican and extends to other branches of Christian orthodoxy, as well as Islam, Hinduism and other faiths.

While some feel genuine horror at the thought of vacuuming a foetus from its mother, others have fewer qualms about early abortions when it is difficult to distinguish the foetus from the uterine tissue that is sucked with it. When the foetus develops limb buds and the brain becomes recognizable, the arguments intensify. Some legal challenges to abortion in the United States have invoked the time when the foetus has a detectable heartbeat; others consider when the foetus can survive in an incubator.

Foetal pain has figured prominently in discussions about abortion and illustrates the complexity of the issues involved.[12] Anatomical studies of early stages of foetal development reveal a bright network of nerve fibres spreading like a miniature river delta from the primordial brain tissues and spinal cord to the extremities of the developing limbs. These are present in six- and seven-week-old embryos in which the different regions of the brain are prefigured as a string of pinhead-sized swellings. Some of these nervous connections deliver sensory information to the brain; others conduct impulses that control movement. Much later, six to seven months into pregnancy, a central part of the brain called the thalamus is hooked up to the cerebral cortex. The thalamus works as a relay station, passing information arriving from sensory nerves distributed over the body to the outer cortex that processes these messages. After birth, these pathways allow us to feel hot and cold, to react to being pushed, to flinch when the skin is cut and so on. Things are more complicated before birth, because it is unclear whether the foetus is ever awake in the womb. Bathed in warm fluid, we are soothed with chemical sedatives that seem to place us in a dream-like state of unconscious awareness.[13] The healthy foetus moves its limbs, responds to loud noises, and kicks and hiccups in the womb, but this does not mean that it is reactive in the conscious way that a newborn baby interacts with its parents.

An adult insect like a fruit fly has a richer capacity for decoding information from its sensory organs and exploring the opportunities and challenges of its environment than a human embryo with an elementary brain structure. Comparisons between insect and human sophistication become less compelling as the pregnancy advances and the most complex brain in nature arises in the unborn baby. Even if the foetus is asleep, it has an astonishingly powerful computer and is receiving information from the womb. What also seems important to me, however, is the unthinking ways in which we mistreat the vivisected and more numerous farmed species of sensitive animals.[14] Beside our unique record of animal abuse, declarations about the unquestionable sanctity of the human foetus loom among the clearest evocations of our stupefying self-regard.

SIX

GENIUS

How We Think

Nerve impulses sprint along neurons at up to 400 km (250 mi.) per hour as we read and think, monitoring our heart and lungs, tensing and relaxing muscles and managing the motion of our gut. Trillions of these signals conduct the mechanical tasks of life, manage incoming information and extract order from a blizzard of thoughts. We are our nervous systems. Although we cannot explain how memories are archived, nor how we draw on them, there is no doubt that the brain accesses stored information about a particular bird, for example, when we think 'bald eagle'. The fact that eagles are recorded in the brain is important for comprehending ourselves. Nobody believes that images of birds come to us from anything other than the brain. By extension, it is evident that human essence – every single thought – lives in the brain and dies with it. Slowly but surely, science has relegated the arguable philosophical concept of the soul to the backwaters of theology.[1] All experience comes to us via the movement of chemicals in our nervous systems.

Human brains are among the most complex computational devices crafted by evolution. Cleverer than cows and chimpanzees, we stride around as self-proclaimed sovereigns and the rest of nature trembles. A good deal of the sophistication of the human brain lies in the outermost layers of its convoluted surface. Our consciousness and power of language come from the signalling between the 16 billion neurons organized in this spongy cap called the neocortex. This structure is a distinctive feature of mammalian anatomy, which makes it a relatively recent innovation. (The term 'neocortex' is often simplified as 'cortex' and this synonym will be followed as we proceed.) Reptiles and older groups of animals cope without this add-on.

Shrew-like creatures called tenrecs that live in Madagascar and parts of mainland Africa are thought to resemble the earliest mammals. Tenrecs have a very meagre cortex that has a smooth surface and is not organized into the clear layers that we can see in slices of human brains. During the evolution of mammals, brain size has increased along some lines of descent and shrunk in others, with the largest brains belonging to elephants and whales.[2] As brain size has increased, the cortex has expanded and developed deep furrows to fit inside the skull. Without cortical folds, our heads would be as wide as extra-large pizzas.[3] In addition to providing humans with unique facilities of language, the cortex handles abstract reasoning and is partitioned into regions that process information from our sense organs. Fine motor skills – like dicing an onion or writing with a pen – are also controlled through nerve cells that lie in the cortex.

It is a mistake, however, to fixate upon the outer brain as the seat of human greatness. Deeper parts of the brain are responsible for a lot of our experiences and eccentricities. Characteristics like risk-taking, fearfulness, baseline levels of optimism versus pessimism, and sex drive are features of our individuality. These behavioural characteristics are controlled by the limbic system, which is built from regions of the brain stuffed beneath the cortex.[4] This antique brain complex is very powerful and we have to be vigilant to keep spills from its ever-fearful chatter offshore. Intelligence is all very well, but a fear of flying or circus clowns can unhinge the best of us; uncensored behaviours and addictive urges from the limbic system can be ruinous, and our inner demon is perfectly capable of conjuring a cloud of gloominess to spoil a sunny day. 'The mind is its own place, and in itself/ Can make a heaven of hell, a hell of heaven', as Satan says in *Paradise Lost* (Book 1, 254–5).

Other parts of the midbrain construct a model of ourselves and our surroundings. As we walk along a woodland path, the brain generates a virtual forest of visual images of trees and shrubs, the sounds of twittering birds and the scent of spring flowers. The actual forest is made from trees, birds and flowers; our simulacrum is created by electrical impulses that arrive from our sense organs and are decoded in the brain. Other animals do the same thing. Moths construct mothy versions of the virtual forest using their array of sense organs and brain centres, which begs the crucial question about the specialness of humans. Over the centuries, philosophers have wrestled with a catalogue of gifts that might define our matchless mentality, from toolmaking and tool use to deceptive behaviour, self-awareness (looking in mirrors) and imitation, only to be outsmarted by the evident accomplishments of other great apes, and monkeys, dolphins, whales and crows. A keen sense of humour seemed like a good bet until we examined the cheerfulness of young chimpanzees and found that they express themselves with screeches that probably qualify as laughter. Dolphins do the same sort of thing, and then came the revelation that playful rats chirp at high frequencies when they are tickled by humans.[5]

Language, which is spoken, written and thought in sentences, is the property that has held up best in this genius contest and there is no doubt that our mastery of verbal communication leaves chimps and gorillas in the dust. Rich vocalizations have evolved among whales, but we have been unable, so far, to translate their speech. Until we can decode humpback and sperm whale conversations, we can feel singularly proud of the linguistic complexity that we employ to share our thoughts, both deep and shallow. Language is the foundation of every other characteristic that seems limited to our species. Returning to the virtual forest, we can choose to imagine that a refrigerator-sized diamond has been placed on the path . . . and there it is! The advent of the impossible gemstone may seem spontaneous, but it begins with an internal conversation. Whales may live wondrous imaginative lives and it seems possible that a bonobo daydreams about mountains of ripe fruit, but the ability to create things that do not exist seems likely to be rare in the rest of nature. Cats stalk birds when they chatter in their dreams, but, without language, they cannot make a deliberate choice to concoct a flock of birds in an empty garden. This language-based facility for fantasy is the source of our artistic impulses, from painting and tattooing our skin and scratching hieroglyphs on rocks, to music and dance, the seascapes of Turner and the poetry of Milton.[6] From language we developed architecture, technology and science. Without it, religion and more complex social rituals are unimaginable.

René Descartes recognized that chin-wagging was our greatest peculiarity and believed that without it other animals were incapable of thinking, at least in the way that humans think. From the observation that humans think about themselves, he developed the doctrine of dualism that separated the mind or soul, which is doing the thinking, from the physical body that is being thought about. Stipulating that souls were only found in humans, Cartesian dualism underscored Christian confidence in our specialness.[7] Dualism has served as a philosophical cornerstone for self-satisfaction at the expense of the rest of nature for centuries. Thomas Hobbes, who met Descartes in 1648, thought that dualism was nonsense, and later philosophers, notably Voltaire, regarded the view of animals as unthinking machines as barbarous. The game is over properly, now that minds have been found in insects.[8]

Inside the bristly head of a housefly, *Musca domestica*, a poppy-seed-sized brain thinks its way through the month of the insect's life. The 100,000 neurons in this little electrical appliance are arranged in distinctive regions dominated by a pair of large optic lobes that plug into the backs of the compound eyes. A lot goes on in the middle of the brain, where a central complex is connected to a pair of stalked structures called mushroom bodies. Mushroom bodies are present in all insects, as well as spiders, millipedes and marine worms. They were discovered in the nineteenth century by a French biologist, who thought that they might allow some insects to impose intelligent control over instinctive behaviour.[9] He based this conclusion on the way that the animals behaved after decapitation, observing that species with the smallest mushroom bodies maintained greatest motor control. More subtle experiments have shown that the mushroom bodies and the surrounding lobes of the central complex serve as a station that directs responses to odours. Mushroom bodies are crucial for learning and memory, allowing insects to recognize smells and move towards or away from them. Mushroom bodies have been compared with the tectum in the human midbrain, which handles information from our eyes and ears, as well as the cortex.

The flying speed of the housefly is a modest 6 km (4 mi.) per hour, but with its wings beating at three hundred times per second, the animal can travel up to 24 km (15 mi.) and land upside down on a ceiling with a deft half-roll. These aerobatics outwit an exasperated human wielding a million-fold-larger brain and a rolled-up newspaper. The insect follows its predator with 4,000 hexagonal lenses in each of its compound eyes, senses the change in air pressure with its antennae, and moves beyond reach one-hundredth of a second before the newspaper slaps the wall. Flies gather more images per second than humans, stretching time so that they have an opportunity to survey the approaching newspaper and take evasive action when its path towards them is certain. Time may seem to pass more slowly for insects.[10] The female mayfly has the briefest life as an adult, during which she is tasked with finding a mate, mating and laying eggs inside five minutes. Her summer holiday is enjoyed in the fifteen seconds she takes to clean her wings and luxuriate in the warmth of the afternoon sun: *carpe secunda*.

Organ pipe mud dauber wasps are beautiful insects whose sapphire bodies and wings sparkle in the summer sunbeams of Ohio afternoons. They patrol individual territories in my garden for weeks, cycling back to the same spots repeatedly to catch and paralyse spiders that serve as live larders for their larvae. The richness of the subjective experiences involved in this hunting behaviour challenges the long-standing view of insects as unthinking robots. The wasp venom robs the spiders of muscular control. The spiders retain their decision-making mechanism but lose their ability to act on the impulse to run away. It is tempting to empathize with the prey and also to imagine the rich mental life of the predator. Perhaps the wasp is joyous when she is pumping her venom into the spider; the spider is almost certainly troubled by its circumstances – he thinks about moving, but his legs do not cooperate. Even if we regard insects as more robotic, or predictable, than ourselves, they certainly offer a model for the wider consciousness that we accord humanity.

Research on insect minds has produced a fearsome challenge to belief in a fundamental difference in consciousness between humans and other animals with brains. Free will is often regarded as a special attribute, but there seems to be no fundamental difference between insect responses to odours and our decisions about whether to pour another glass of wine or recork the bottle. German philosopher Arthur Schopenhauer understood that free will was an illusion: 'You can do what you will, but in any given moment of your life you can *will* only one definite thing and absolutely nothing other than that one thing'.[11] The choices that we make in response to particular opportunities and challenges are not as unencumbered as we believe. With shorter psychiatric repertoires, the same applies to insects and worms. Wilfulness in worms was established by examining their response to a pungent scent found in truffles.[12] The activity of three nerve cells that form a circuit in the brain determines how the worm responds when it detects the odour. Sometimes, the worm changes direction and wiggles towards the odour, but it may continue on its path if it is occupied with something else. By modifying the activity of the trio of nerves, researchers were able to control the reactions to the truffle smell, robbing the worm of choice.

When we look at the behaviour of insects and worms, it is clear that their responses to odours and other stimuli are probabilistic, meaning that there is a statistical likelihood associated with a particular reaction. This goes beyond the behaviour of a robot vacuum cleaner that is programmed to turn around every time it hits an obstacle. More intelligent robots are like these simple animals and equipped with probabilistic software. Once we strip our notions of consciousness and free will to their essence, it is clear that all life, and every individual cell, expresses some version of decision-making in its activities.[13]

Slime moulds grow on rotting wood and can be taught that sliming in one direction, rather than another, will be rewarded with food. Without a brain, with no nerves of any kind, these microorganisms respond to training like dogs to Pavlov's bell and humans to the smell of roasting coffee. There is a single-celled alga that lives in the sea and is equipped with an eye with a lens and a retina that it uses to spy on shadows thrown by its predators and prey. Like a multi-cellular organism, this amazing microbe gathers visual images and responds by swimming away from a threat and towards its food. And, in a final example of non-nervous sensitivity, a fungal colony can form 130 billion microscopic threads in a kilogram of soil, allowing chemical communications across the colony as it searches for food scraps and connects with plant roots. The fungus alters its growth when one segment of the colony signals that it has located a meal, and also shuttles nutrients to plants that express their willingness to cooperate.[14] All biology is feeling, thinking and talking.

While other animal brains operate according to the same principles as the human brain, ours is very big for an animal of our body size. The walnut-sized brain of a cat represents a little less than 1 per cent of its body mass. Our cantaloupe-sized brain represents 2 per cent of our carcass and devours one-fifth of our calorie intake. Larger mammals tend to have bigger brains than smaller mammals, and if we take this into account we would expect the human brain to be the size of an orange. This may seem inadequate to the task of being a human, which it is, but a little brain like this did work for one of our hominid relatives, named *Homo naledi*, that lived in South Africa.[15]

The brain sizes of other close relatives, which we have gauged from the dimensions of their fossilized skulls, have varied a good deal, from half a litre in australopithecines, to a whole litre in *Homo erectus*, and more than one and a half litres in the Neanderthal. Multiple species of australopithecines, or australopiths, spread over East Africa 3.5 million years ago; *Homo erectus* evolved from australopiths and migrated eastwards as far as China. Neanderthals, which some palaeontologists regard as a subspecies of *Homo sapiens*, evolved 250,000 years ago, were concentrated in Europe and became extinct 40,000 years ago. Neanderthals were stronger than us, had better eyesight, ate more food and grew slightly larger brains. Shared genes suggest we mated with these robust relatives.[16] Every misbegotten claim about human superiority comes back to the remarkable increase in brain size that occurred in a twinkling of time in the skulls of bipedal apes.

We are not sure why genes that increased brain size were rewarded in our history, but there are a number of compelling theories. Natural selection sorts effective from ineffective genes, granting life to the best kinds of vehicles and dispensing with the rubbish. It follows that larger brains allowed our ancestors to leave more offspring. The most likely explanation does not speak well of us, and Thomas Hobbes said it first and said it best: 'the state of men without civill society (which state we may properly call the state of nature) is nothing else but a meere warre of all against all', which he expressed in his famous Latin aphorism, *bellum omnium contra omnes*.[17] Big brains were an asset for dodging our predators and hunting other animals, and even bigger brains made us successful in combat with other hominids. Our intelligence evolved as a weapon.[18]

Our prehistoric victories in this struggle against the hairier majority of the hominid family are undeniable. Slaughter ensued as we clubbed and speared the species that were unfortunate enough to meet us on the grasslands, and changing environmental conditions put an end to the others. Over thousands of millennia, apes have come and apes have mostly gone and we are the last of our genus. We exterminated the Neanderthals in Europe and our small-brained relations like the hobbit, *Homo floresiensis*, in Indonesia.[19] Extinction of these species follows the broader pattern of large animal disappearance whenever humans show up. Violence against competitors is a matter of survival, but we have a durable record of killing for sport too. Violence within the species, or tribal warfare, is another skill that we have perfected and is likely to have caused the exodus of groups large and small from Africa as the human population mushroomed and ran out of food and water.

Competing theories suggest that sexual selection has come into play in our history, with females choosing males with bigger brains. The idea here is that big brains are associated with artistic talents, storytelling, dancing or some other characteristic that females find appealing. Sexual selection can act swiftly, driving brain expansion in a runaway fashion reminiscent of the presumed origin of peacock tails and deer antlers. Other biologists are committed to the possibility that brain size increased as our social interactions deepened, from the necessity of collaborative hunting to the adoption of specialized roles like tool manufacture and cooking. The likeliest explanation for our smarts lies in a combination of these factors, because the brain required of a great military tactician is likely to be good at other things. Julius Caesar and Ulysses S. Grant were masters of the pen and the sword, Joan of Arc danced like an angel and Napoleon played chess.[20]

As bearers of the most powerful intellects in the universe – at least in the immediate galactic neighbourhood – we gaze at the night sky and feel compressed by the immensity of time between these points of light. No other creature can ponder these things. In this sense, it seems fortunate to be human rather than insect, but big brains are an encumbrance too, telling us that consciousness is short-lived and that all these moments will be lost in time, like tears in rain. The limbic wiring sends reminders to hardcore thanato-phobes like myself with appalling regularity. It is with some unease, then, that I dedicate the next chapter to the loss of life and how it is achieved. Perhaps there is a silver lining?

SEVEN

GRAVES

How We Die

'One by one, the lights go out and there is total blackness,' Christopher Isherwood wrote of death in his novel *A Single Man*.[1] We all know that this is forthcoming, but the reason for death can be difficult to grasp. The Bible explains that death was God's punishment for Eve's decision to favour knowledge over obedience. And, whatever we believe about the myth of Eden, mortality certainly feels like a serious punishment when we are enjoying life. The only reliably good thing that people say about death is that it clears the field for the next generation: grandparents need to shuffle off to make way for grandchildren.[2] What we fail to recognize with this comforting reflection is that funerals for children are a more effective method for stemming population growth than funerals for the elderly. This leaves us in a quandary. If the demise of the elderly is not required to make room, why do we have to become rickety, witness the ruination of our dearest friends and, ready or not, race or shuffle towards the exit ourselves? If we are not doing this for charitable reasons, perhaps we can refuse to go? Christopher Marlowe's Dr Faustus thought this might be possible, but then the devils enter the final scene of his play and he says with awful resignation, 'Ah, Mephistopheles!' – and off he goes like everyone else.[3]

There is a lot of fuzzy thinking about the reason that we age and die, and the puzzle was not solved until biologists embraced a clear, gene's-eye view of ageing in the middle of the last century.[4] Here is the answer: animals are mere vessels for genes. Effective genes permit individual animals to survive, which makes it more likely that they, the *genes*, will be passed on to future generations. Evolution is blind to the failings of older bodies. Death comes because there is no biological value in the prevention of decrepitude.[5] Our bodily mechanisms have been forged for survival until testes descend, ovaries start popping, and we have foisted our genes upon fresh humans.[6] There is no demand for wrinkly people to keep mating like rabbits when the youth are so effective in this department. But mortality does not possess any inherent evolutionary advantage, which makes it unlikely that there are death genes purposed with killing us.[7]

Wrinkles and other expressions of diminishing elasticity come from molecular changes in our cells. These include the formation of wonky protein molecules and errors in the quality control mechanisms that get rid of these misfits. A related problem is that caps at the ends of chromosomes that protect genes become shortened each time a cell divides. Cells deteriorate as this trimming proceeds, the immune system becomes sluggish and age-related diseases set in. Exhausted by protein problems and shortened chromosomes, cells are further disabled by the discharge of reactive chemicals from mitochondria and the ageing nucleus bloats like a dying star.[8]

Ageing is inevitable because evolution concentrates on the genetic programmes that shape bodies from fertilized eggs into adults that see to fertilizing the next generation of eggs. The accumulation of faulty molecules in geriatric cells is an illustration of the relentless increase in disorder, or entropy, that pervades the universe.[9] The entropy rule is expressed in the Second Law of Thermodynamics:

$$\Delta S = \delta Q / T$$

where ΔS = change in entropy, δQ = heat transfer and T = temperature. This dictates that, sooner rather than later, our warmth will be transferred (δQ) into the environment and we will not be detectable from the background temperature (T). Hold an old-style thermometer and watch the mercury rise; put the thermometer on a table and watch the mercury fall. 'I am alive—I guess', wrote Emily Dickinson.[10] In his last months, the writer Christopher Hitchens described his entropic sensation of 'dissolving in powerlessness like a sugar lump in water'.[11]

Above the grave, our bodies are islands of orderly molecules cast within a cooling universe. Entropy is evident from the accumulation of errors in gene expression, as well as in the incessant damage caused by viruses and a plethora of environmental poisons.[12] This is why the maximum lifespan is never likely to creep more than months beyond the purported record of 122 years and that few of us will enjoy more than 33,000 days. We live longer than most animals with backbones, including an Australian fish that wraps up the whole business in two months, but we fall short of the Greenland shark that may live beyond its 300th birthday.[13] If we bring invertebrates into the comparison, we may feel further elevated by the demise of a nematode worm after only three days of wriggling, and disheartened by an Icelandic clam that persists in its clamminess for more than 500 years.[14]

Encouraged by sharks and clams, devotees of life extension embrace hormone replacement therapies, vitamins, enzymes, antiviral drugs, fish oils, plant extracts and dried mushrooms. Traditional Chinese medicine is replete with miracle cures for mortality that come at the expense of endangered species, yet no amount of powdered rhinoceros horn or pangolin scales keeps anyone from the grave. With similar vanity, the Californian fetish of deep-freezing heads seems as likely to work as removing organs from a corpse, wrapping it in linen strips and stowing it in a pyramid. Even the most enthusiastic backers of this gelid pursuit seem sceptical about the outcome, because none of them elect to have their head submerged in liquid nitrogen *before* they are dead.

The fear of the process of dying is more reasonable, perhaps, than the fear of its aftermath. Everyone would prefer to sidestep a painful death, and even when expiration is embraced as a relief from pain most feel sad when they think about missing the amusements of the future. The Roman poet and Epicurean philosopher Lucretius offered some hope with his cheerful insistence on the symmetry between being unborn and being dead, reminding us that we have spent a great stretch of time in the former state already.[15] This is all very well, but modern pharaohs pursue reincarnation with all of the vigour of ancient royalty and place their faith in an *in silico* persistence after their brains are uploaded to computers. Feasibility studies on the pseudoscience of Whole Brain Emulation suggest that each brain requires a computer with one petabyte of memory (10^{15} or one quadrillion bytes).[16] This is a problem, at least at the time of writing, because the largest single-memory computer, which is manufactured by Hewlett Packard Enterprise, has a measly 160 terabytes of memory (six times less than a brain). It appears that we are too smart to be copied, which seems a remarkably positive evaluation of our brainpower when we can forget the name of someone of nagging familiarity at a committee meeting. The explanation lies in the difference between the storage capacity of a brain and its processing speed. Nerve impulses are transmitted very slowly compared with the gigahertz clock speeds of a computer chip, which explains why the person's identity is revealed as the meeting crawls on.

The range of technical obstacles confronting plans to duplicate brains is immense and we will not be in any position to emulate a fruit fly, let alone a Californian virtuoso, until we understand how information is stored in brain cells. Even if we solve these issues and create a functioning brain facsimile, the experiences of this synthetic doppelgänger would be completely separate from the past life of its blueprint. It could turn angel or monster. Consider this: does the dead sibling of an identical twin live on in its surviving brother or sister? In a poetic sense, surely, but not in a way that concerns the deceased. Two thousand years after Lucretius, fantasies about immortality persist as another of the great acts of self-indulgence characteristic of *Homo narcissus*.

Prospects for immortality improve elsewhere on the tree of life. Microorganisms mastered the craft a long time ago. When plenty of sugar is available, a yeast cell copies its chromosomes and pushes one of the sets into a bud that grows from its surface. A single yeast, or mother cell, can divide in this fashion for up to a week, producing twenty buds, or daughter cells, before she is disabled by malfunctions in gene expression and protein recycling. Meanwhile, each daughter goes on to produce its own family of buds. There is a marvellous rejuvenation process at work here, which excludes all of the geriatric flaws of the mother cell from her buds.[17] This allows yeasts to obtain a sort of immortality. Individuals die but their genomes live on in their buds. Mammals cannot do this. The best that we can do is transmit half of our genes into a baby, one-quarter into a grandchild, and have our personal arrangement of DNA jumbled beyond recognition within a few generations.

The most complex animals that have a claim on immortality are jellyfish. Jellyfish are remodelled through their life cycles, from tiny swimming larvae to colonies planted on the seafloor that shed the familiar pulsating bells with trailing tentacles. The bells produce eggs and sperm cells and the fertilized eggs turn into the larvae. A few species of these sea creatures have the remarkable facility of shrinking, retracting their tentacles and reverting to life as an attached colony.[18] This trick may be likened to butterflies turning back into larvae, or residents of a retirement community awakening as children. Jellyfish rejuvenation was discovered in animals reared in jars of seawater and it is not known how often it occurs in nature. Sex continues to be the most effective means of reproduction, however, because most jellyfish die from old age or are eaten by predators in the sea.

Experts in the field of regenerative medicine, which concerns the revitalization of damaged cells and failing organs, are excited by the reversal of the jellyfish life cycle. This developmental plasticity encourages the idea that we might be able to replace worn-out body parts with fresh tissue. Experimental therapies using human stem cells are the most encouraging part of this endeavour. Stem cells are cells without a specific career plan that have the potential to take on a particular role in the future. All of the two hundred kinds of cell in the human body come from the division of the fertilized egg. Complete flexibility is retained by the cells that form the ball of the blastocyst stage of the embryo. This explains why embryonic cells are so valuable from a medical perspective, but their use to treat illnesses raises some rather important ethical questions. Stem cells in bone marrow are an alternative, but their developmental options are limited to becoming different kinds of blood cell. Blood drained from an umbilical cord or placenta is another less controversial source of stem cells that can be used to treat blood disorders. Stem cell therapies show great promise for treating a range of otherwise fatal diseases, but they seem as unlikely as head-freezing to keep us running beyond the maximum lifespan rooted in the Second Law. 'Golden lads and girls all must,/ As chimney-sweepers come to dust,' as Shakespeare wrote for the funeral song in *Cymbeline* (Act IV, Scene 2).[19]

How will you die? Pump stoppage after three billion contractions is most likely, with organ damage by cancer cells coming a close second. Collapse of the breathing apparatus due to chronic obstructive pulmonary disease is the third commonest swansong. Together, these causes of poorliness end the amusements for half of the citizens of developed countries. Accidents and strokes take another one in ten, and the rest of our terminal complaints serve as deaths privileged, speaking statistically, for the select few that expire from the diagnoses named for Drs Alois Alzheimer (whose heart failed when he was 51) and James Parkinson (who stroked and died at 69), and sundry vulgarities of the liver and kidneys.

Infectious diseases were a reliable death sentence before physicians saw the benefits of science. With attention to public sanitation, the introduction of hand-washing by obstetricians and surgeons, and the invention of vaccines and antibiotics, we began to survive long enough for our heart muscle to get tired and other cells to lose their brakes and fill us with tumours. Where expensive drugs and clean water are scarce, the microbes that cause AIDS, pneumonia, diarrhoea and malaria continue to waste populations in poor countries. Death by explosive devices is an additional risk in centres of internecine and international conflict. One way or another, the goddess Fortuna spins her *Rota fortunae*, or Wheel of Fortune, and we are swept away. More than 1 per cent of humans take control of the situation by noose, gunshot, poison and jumping. Greenlanders top the suicide charts, with one-quarter of the denizens of this wintry isle attempting to kill themselves at some juncture in their darkened lives.[20]

Greenlanders and the rest of us do not die when we imagine that we do, but keep chugging along, genetically, for many hours after the last exhalation. A Spanish study of fresh cadavers found that genes encoding the formation of heart muscle remained active, along with protective genes that control inflammation.[21] It seems that the body responds to the drop in oxygen levels by trying to resuscitate the heart. This interpretation is supported by the post-mortem expression of another set of genes that shape the heart in the developing embryo. These genes are turned off once we escape the waters of the womb, and their reactivation after cardiac arrest shows that the body reaches deep into its toolkit to get going again. In the midst of death, it seems, we are in life: 'we have some salt of our youth in us'.[22]

Even as the failed heart is being patched, there is an uprising in the gut. Onboard microbes that need oxygen are gasping for breath and the entire microbiome is wondering what happened to the meal service. The suffocated cells of the intestinal wall begin to leak, providing a feast for the bacteria that are less fussy about oxygen. Without an immune system there is no policing of the microbiome, and the microscopic peasantry pours through the barricades, eating us from the inside out.[23] Internal and external microbes, followed by insects and worms, digest every scrap of soft tissue. Whiskered rodents nibble at the tougher bits and birds drop down to peck and pull. Exposed by tooth and beak, the bones lie in the wake of this fervent composting and embark upon their slow dissolution into the damp earth.

There is a view of decay that may seem less regrettable when we dwell on our cosmic brevity and consider future months 'a-mouldering in the grave', without the prospect of a soul that marches on. This involves the difficult task of dampening our egotistic impulses and trying to connect with the truth that we live as ecosystems rather than as single entities. You and I, mortal mammals, are collaborative ventures between human and bacterial cells, in which we share food and chemical signals and rely on the immune system to serve as ringmaster of the gut circus. Even appetite and mood are affected by a continuous chemical conversation between microbe and man. We have no physical awareness of our communal nature, but what feels like *me* is *us*: *Cogitamus ergo sum* – we think, therefore I am.[24] This is crystal clear for a Buddhist and the Qur'an says as much for Muslims: 'all creatures that crawl on the earth and those that fly with their wings are communities like yourselves' (6:38). And when we die, all the stuff that allowed us to care about being, to love and to feel loved, and, with great fortune, to glimpse the reasons we are here by turning a microscope and a telescope on this scrap of space, seeps back to its source.

Having quoted from Isherwood, Marlowe and Hitchens on these grave issues, we close with a rosier sentiment from a fourth notable Christopher:

> Christopher Robin came down from the Forest to the bridge, feeling all sunny and careless, and just as if twice nineteen didn't matter a bit, as it didn't on such a happy afternoon, and he thought that if he stood on the bottom rail of the bridge, and leant over, and watched the river slipping slowly away beneath him, then he would suddenly know everything that there was to be known, and he would be able to tell Pooh, who wasn't quite sure about some of it.[25]

EIGHT

GREATNESS

How We Make Things Better

On the same day in 2016 that two suicide bombers killed 58 Nigerians in a camp established as a safe haven from terrorist attacks, an international team of physicists announced that they had detected gravitational waves produced by the collision of a pair of black holes.[1] Pick a day, any day, and science will provide a modest counterbalance to our abundant shortcomings as a species. Even when the practical benefits of a discovery are unclear, or non-existent, we can bask in the glow of each epiphany about nature. *Homo narcissus* obtains a state of grace in its scientific breakthroughs. Carl Sagan wrote that 'Science is more than a body of knowledge, it's a way of thinking,' which explains why everyone deserves to learn how it works.[2]

Francis Bacon, who framed the experimental method of modern science in the seventeenth century, was less sentimental than Sagan about the beauty of science. He believed that knowledge should not be pursued 'for pleasure of the mind, or for contention, or for superiority to others, or for profit, or fame, or power, or any of these inferior things; but for the benefit and use of life'.[3] Bacon was frustrated by the petty pace of scientific progress in his time, which he blamed on the enduring authority of Aristotle, who had 'made his natural philosophy completely subservient to his logic, and thus rendered it little more than useless and disputatious'.[4] Aristotle had argued that careful thinking and educated guesswork led us to the truth. Experience tells us that this deductive method can be very effective, but it has the considerable drawback of leading us in the wrong direction if we make a single faulty assumption. Bacon wanted humanity to grow up fast and championed the process of induction, which sought answers after gathering many facts by performing experiments.

Western science is on display throughout this chapter – and throughout the whole book. No correction is justified for such hegemonic absorption. It would be disingenuous to concentrate on Sumerian agriculture, Persian astronomy and Chinese chemistry when the application of Baconian methods has brought us so far. Gender privilege is on parade too, with men dominating science for a range of indefensible reasons. With these caveats in mind, humans have made some splendid scientific discoveries in the last four hundred years. Galileo Galilei humbled Earth as a satellite and Isaac Newton figured out how and why it moves around the Sun; Robert Hooke startled plague-ravaged London with gigantic illustrations of fleas and lice; Charles Darwin shocked Victorians with his evocation of natural selection; in the twentieth century, Albert Einstein transformed physics with his claim that time and space are the same thing. But another discovery, from the 1950s, looms larger in my imagination. This is the structure of DNA.

Much of the story is familiar. Watson and Crick solved the arrangement of the different chemicals in the DNA molecule by building models using cut-outs of the natural components made from cardboard and sheet metal.[5] The famous metal model of the threedimensional structure was assembled after Watson had an epiphany about the arrangement of the chemical groups inside the double helix by shuffling the cardboard pieces around on his desk. Vital clues to the shape of DNA came from experiments by Rosalind Franklin, who bombarded DNA fibres with X-rays and captured the scatter patterns that they made on photographic film. Watson used the information gathered by Franklin without her knowledge and he has been vilified by some biographers for failing to acknowledge the importance of her work.[6] To explain his eagerness, without defending his lack of decency, he was an ambitious 24-year-old engaged in fierce competition with other very formidable scientists who were hell-bent on making the next breakthrough in the emerging field of molecular biology. Someone was going to crack the structure of DNA very soon and win a Nobel Prize.

The DNA story is made more interesting by the failure of Linus Pauling to get there first, and work by a group of more admirable scientists at University College Nottingham who had come close to solving the puzzle a decade earlier. Pauling was an expert on the bonds that hold atoms together in molecules and had discovered the way that proteins are folded and twisted into shapes that are crucial for their function. But when it came to DNA, he made a series of fundamental mistakes. The best way to explain this is to consider the actual structure of DNA. DNA is coiled into a spiral staircase with steps between the outer rails. DNA stands for deoxyribonucleic acid. The sugar molecules – deoxyribose sugars – in the rails are connected by groups of oxygen atoms that release positively charged hydrogen atoms, or protons (H^+), when they are bathed in water. This process leaves negative electrical charges on the outside of the molecule, which is what acids do. Pauling imagined that DNA had three rails and that these were buried on the inside of the molecule, with the rungs split in half and forming pegs that faced outwards. Pauling's DNA looked like a toilet brush. One of the problems with this phantom molecule is that it would not work like an acid, and in any case, the negative charges would resist being stuffed into the middle of this triple helix, repel one another and blow the structure apart. But Pauling was so desperate for the fame that would come from solving DNA that he grabbed at this inside-out answer in 1953.[7] It is worth mentioning that he received the Nobel Prize in the same year for his work on chemical bonds. He was besotted by his own genius and believed he could do no wrong.

The chemists at University College Nottingham were far quieter in their contributions to science. In the 1940s they proposed that DNA had two strands that were held together by a special type of bond between the pairs of bases that formed the rungs in the middle of the molecule. Evidence for this model came from experiments in which they raised or lowered the acidity of mixtures of purified DNA and found that the molecules fell apart. Bonds that behave in this way are formed by hydrogen atoms. The youngest member of the team, Michael Creeth, drew a diagram of DNA as a straight ladder, rather than a helix.[8] He was eerily close to the truth. If Linus Pauling had met with Creeth and his colleagues when he visited their campus in 1948, he might have avoided his crooked rendering of DNA.

Watson had seen the experiments published by the Nottingham chemists, but did not recognize their significance at first. As his investigations on DNA matured, he read their papers again, realized his mistake and conceived the correct arrangement of the building blocks with Crick in a few days. They published the structure of DNA in 1953, and were awarded the Nobel Prize in 1962, along with Maurice Wilkins, who was Rosalind Franklin's supervisor. No more than three scientists can share a Nobel Prize, so one wonders whether Franklin would have replaced Wilkins on the stage in Stockholm if she had not died from ovarian cancer in 1958.[9]

The experiments by Watson and Crick were very limited, in the sense that they did not require any complicated and time-consuming techniques and were completed in a few weeks. The dynamic duo were quite Aristotelian in their dependence on educated guesswork. But if we look at the whole story of DNA, rather than the race at the end, it is clear how the method of induction played out. Watson and Crick relied on information gathered by many other investigators who contributed snippets of crucial data that fed into the right answer. Indeed, the study of DNA began a century earlier, with the work of Friedrich Miescher, a Swiss chemist who isolated a mixture of nucleic acid and protein from pus cells that he washed from bandages.[10]

DNA is a stunningly beautiful molecule. To act as the vehicle for transmitting information across billions of years through an infinitude of living things, it had to possess such gorgeous symmetry. The two strands are complimentary, bearing the same information, so that by pulling the thing apart, each single strand serves as a perfect template for making another strand. This duplication of the instruction book – written in 3 billion As, Ts, Gs and Cs in our genome – is essential every time a cell divides. When Watson and Crick looked at their model, they saw, immediately, how DNA was copied.

Their discovery of the double helix is one of the greatest scientific accomplishments in history. There is a good deal of self-indulgent speciesism in this assessment, because it laid human nature bare and the resulting molecular technology is changing the way of medicine. Other scientific advances are every bit as impressive, but they have not affected us in the same way. Breakthroughs in cosmology like the detection of gravity waves are very exciting, but they do not say anything specific about humans. Indeed there is a sense in which each revelation by astrophysicists belittles us as it enlarges the magnificence of the universe.

Gregor Mendel demonstrated that characteristics of organisms were passed from generation to generation in the form of units of information. Crosses between pea plants of normal height and dwarf plants generated seeds that produced a predictable ratio of normal and dwarf plants when they germinated. He had no way of knowing how chemicals carried these instructions, only that something exchanged between parent plants had this property of affecting the growth of their offspring. Mendel's mystery melted once the structure of DNA was disentangled and geneticists began to understand how genes worked. With the knowledge of genes came a clear view of the way that their alteration by mutation served as the raw material for evolutionary innovation.

These paths of exploration led scientists to answer many of the big questions in biology in the second half of the twentieth century. With the unmasking of genes, scientists finally became clever enough to offer a comprehensive understanding of life. We are the beneficiaries of this scientific quest. The double helix calls to us. This is what I am, nothing more, nothing less; beauty and beast, all wrapped up in a spiral staircase.

The practical implications of the DNA discovery are expanding today. With the ability to sequence genes, to mutate them in the lab and to transfer DNA between species, biotechnologists have transformed microorganisms into industrial tools that manufacture powerful drugs. We can pinpoint the causes of a whole range of inherited diseases and make predictions about the likelihood that chronic illnesses will develop from the presence or absence of different versions of genes. The techniques of molecular genetics can also be turned to establishing our ancestry, settling paternity disputes and tracking criminal activity by amplifying traces of DNA. Watson and Crick played no direct role in these advances, but we would be stuck in the mud without the discovery of the double helix.

The re-engineering of microbes with human genes illustrates the undeniable power that has come from understanding DNA. Insulin was the first protein manufactured by bacteria and yeast modified with a human gene. Insulin is a simple protein, as proteins go, which is built from two chains that twirl along portions of their length and are tethered by chemical bridges. This structure was solved by Dorothy Hodgkin, another brilliant scientist who happened to have a pair of X chromosomes.[11] Like Rosalind Franklin, she used X-ray crystallography to anatomize biomolecules. It took her thirty years to make sense of insulin crystals – so long that she received a Nobel Prize for other work as this research played out. Insulin has none of the symmetrical sexiness of DNA, but without this molecular glob we cannot absorb sugars from the bloodstream and our cells stall for lack of nourishment. Before we figured out how to inject diabetic patients with insulin extracted from pigs, sufferers were blinded, lost limbs, had strokes and heart attacks and died from kidney failure. Starvation diets flattened some of the symptoms of diabetes and offered patients an extra year or two of misery; opium served as a palliative; there were no other treatment options. The use of porcine insulin to control blood sugar began in the 1920s, but once the human insulin gene was sequenced, it was not long before bacteria and yeast were engineered to make it for us. This works because microbe DNA is written in the same A, T, G, C alphabet as human DNA, and bacteria and yeast use the same genetic code and decoding apparatus to translate these letters into proteins.

The ultimate goal of molecular medicine is to wipe out illnesses whose causes lie in our mangled genomes. The approach sounds simple: replace defective genes with flawless DNA sequences and the illness will be fixed by the production of the healthy proteins. We are three decades into this grand experiment. Gene therapies are advancing very swiftly, with pharmaceutical companies exploring treatments for muscular dystrophy, cystic fibrosis, bladder cancer, cervical cancer and a range of inherited diseases. In 1989 investigators identified a gene, *CFTR*, whose mutated sequence caused cystic fibrosis.[12] *CFTR* encodes a protein that controls the movement of chloride ions in and out of cells. Mutations in the gene stop this flow of ions, which causes thickened mucus to accumulate in the lungs. This was the first time that a gene that caused a specific illness had been characterized. A cure for cystic fibrosis seemed within reach; all we had to do was fix a single gene. It has not proven that simple.

The most promising gene therapies use a virus to import the corrected version of the target gene into cells and provide enough of the healthy version of the protein to treat the condition. Cystic fibrosis has resisted this remedy because the mucus in the lungs acts as a barrier against the uptake of the virus. Other challenges include the continuous replacement of the cells that line the lungs, which means that patients have to be given repeated doses of the genetically modified virus. Finally, the *CFTR* gene is expressed in all tissues, not just the lungs, which explains why cystic fibrosis patients can experience problems in other organs. Prospects for treating haemophilia are significantly brighter.[13] Trials using viruses carrying human genes that code for the missing proteins that coagulate blood have shown impressive results, with swift wound repair in adult patients who have suffered from uncontrolled bleeding since childhood.

The promise of new disease treatments is a perfect example of the manipulation of nature 'for the benefit and use of life' (meaning human lives), just as Bacon intended. Four hundred years after Bacon, this 'practical applications' argument remains the typical plea for funding science. Shallower politicians, especially in the United States, accept this when the title of a research project refers to a childhood illness, then deride experiments on fruit flies without recognizing their importance as models for human disease. They mistrust what they do not understand. The provision of funding for investigations on flies for their own sake, without claiming that there is any human benefit, is a non-starter for the majority of taxpayers. For this reason, scientists make splendid arguments about the way that their work may improve the lot of *Homo narcissus*.

The more ardent defenders of scientific research recommend that all of it should be funded, because we cannot predict where the next breakthrough may lie. It is true that it is difficult to forecast major advances, but we can be pretty confident that many areas of inquiry are dead ends that will never tell us anything useful or even very interesting. Many hopeless pursuits are evident in my own field of fungal biology. Thoughtful entomologists and particle physicists will admit the same thing about their areas of specialization with a little persuasion, but we tend to reserve our criticism for the anonymous peer review of publications and grant proposals. James Watson said it very well:

> One could not be a successful scientist without realizing that, in contrast to the popular conception supported by newspapers and mothers of scientists, a goodly number of scientists are not only narrow-minded and dull, but also just stupid.[14]

Even when we accept the shortcomings of scientists – always *other* scientists, of course – the complex sociology of the endeavour, with its guilds of academy members, durable misogyny, and insistence on the facade of politeness, makes it difficult to ensure that the best research gets the attention it deserves. Despite all of these warts, Western science has served as a beacon of human greatness since the Renaissance. It is the activity that we will proclaim in our defence as an intelligent animal when the alien auditors visit from a nearby solar system. Poetry and music will be next on the list. But what if the whole scientific adventure has been the fatal error of our species, the source of the very technology that will destroy civilization?

NINE

GREENHOUSE

How We Make Things Worse

The decline of our species was the natural and inevitable effect of immoderate greatness. The story of its ruin is simple and obvious; and, instead of inquiring why humanity was destroyed, we should rather be surprised that it had subsisted so long.[1]

That we have created conditions on Earth that will hasten our decline is undeniable. Here are the circumstances: Earth is warming swiftly; seawater is acidifying and choking with plastics; industrial activity is poisoning the air; deforestation is relentless; grasslands and lakes are shrinking as deserts expand and a swarm of 10 billion humans will jostle for the remaining resources by 2050.[2] In short order, extreme weather events will become more frequent; crops will be withered by drought; fisheries will crash; populations of the larger wild animals will continue to dwindle; insect numbers will pursue their precipitous decline; plant species will perish and the microbial majority of life will shudder unseen.[3] On a somewhat longer timescale, coastlines will be reshaped by rising sea levels.[4] As the Antarctic ice sheet calves and dissolves, Florida and Bangladesh will vanish beneath the waves. These planetary changes may be imperceptible to you so far and it seems plausible that your circumstances will be sustained in the immediate decades. After all, wealth is the most reliable cushion against many of the exigencies of life. But even the nobility should consider the ecological future before having children.

The unfolding story of Earth's ruination has involved some outstandingly maleficent corporations, yet everyone is culpable and the climatic apocalypse was stamped in our genes from the moment we disgorged from the Rift Valley.[5] We share our impulses for eating and mating with mice and mushrooms, but unlike other organisms, the misfortune of brain power has allowed us to feed and breed in ever-increasing numbers. Besides the environmental impact of the head count, the luxuries of modern life multiply the planetary damage. Most people want to live like royalty, and as opportunities present themselves, we have an understandable tendency to make life more comfortable. These perks have come at the expense of the gas composition of the atmosphere, thickening the blanket of carbon dioxide and trapping the warmth of the sun on Earth. It is not possible to be sure how far we will roast the planet, nor how fast we will heat up, but warmer we are getting.

My Texan brother-in-law is untroubled by the evidence. He cites the Medieval Warm Period and finds solace in the writings of sundry deviants who deny the startling correspondence between carbon dioxide emissions and average temperatures. His views are widespread in the United States, where white citizens in particular are unaccustomed to the idea that 'life, liberty, and the pursuit of happiness' might be deniable. In many other parts of the world, the cause of hotter summers is ignored by people who are too busy trying to survive to worry about invisible gases.

I write this with deep humility, as a contributor to the end of civilization, driving short distances rather than bicycling, flying internationally and purchasing strawberries from South America in indestructible plastic containers. I will not willingly live in a tent, but in my defence, I probably have a smaller carbon footprint than most of my neighbours. As a step-parent, rather than a biological one, I would have to commute to work in a coal-burning Lear jet to do as much damage as anyone who has donated sperm or eggs to the next generation.[6] The greatest contribution that an individual can make to reducing greenhouse gas emissions is to be dead. Failing this, the next best thing is to abstain from manufacturing babies.

Thomas Malthus was the first to recognize the peril of unmitigated human replication in his *Essay on the Principle of Population*, which was published at the dawn of the Industrial Revolution.[7] His interest lay in the potential for mass starvation attending a geometrical increase in the number of mouths. This diagnosis of the human condition was corroborated by the Irish Potato Famine in the 1840s, but land development, the introduction of fertilizers, herbicides, pesticides and the mechanization of agriculture – all reliant on fossil fuels – furnished us with a false sense of security in the twentieth century. Combined with advances in medicine, the boom in farm yields has allowed the human population to quadruple in the last hundred years.

The relationship between population growth and environmental degradation is neglected in public discourse. Politicians never engage with this topic, and the kind of doomsday scenario raised by Paul R. Ehrlich in his 1968 best-seller *The Population Bomb* is regarded by the majority of public intellectuals as lunacy.[8] Contemporary economists are more concerned about the falling populations of developed nations than the soaring numbers in other parts of the world. Even the most luminous environmental activists ignore population in their pronouncements on sustainability, as well as in their personal behaviour. Al Gore, 45th Vice President of the United States, fathered four children, and fellow politician and activist Robert F. Kennedy, Jr, added six children to his fabled clan. Rather than a badge of honour, having lots of children in the twenty-first century is an act of environmental terrorism. Fifteen thousand children are born every hour and only 6,000 people die; the sums do not favour the future.

Humans are not the only organisms to have affected the liveability of Earth. Microbes and plants changed the chemistry of the atmosphere long before we leapt on to the stage. Bacteria initiated a momentous change 2.3 billion years ago when they began flooding the air with a noxious gas called oxygen. Microorganisms that had been happily 'breathing' iron, sulphur and nitrogen for the first million millennia of biology were decimated by this highly reactive, DNA-damaging molecule. As oxygen levels rose, the metal-breathers and their kin retreated to marine muds and other oxygen-free quarters. New life forms evolved to take advantage of the peculiar conditions and found a way to use oxygen to rip more energy from their food, which is why we breathe deeply today.

Much later, after life had crawled on to the land, the gases in the air shifted again in accordance with the abundance of plants. Giant horsetails and club mosses that rioted in the luxuriant forests of the Carboniferous Period were spared decomposition and became pressed into seams of coal.[9] This custom of burial without decay was so effective at drawing carbon dioxide out of the air that it caused global cooling. We unlock this carbon every time we draw electricity from a power station that burns coal, to send photons quivering from light bulbs at the same wavelengths that were absorbed by the prehistoric forests. Energy in and energy out, from fossilized greenery to table lamps after more than 300 million years. Coal formation declined after the Carboniferous, when fungi mastered the decomposition of fallen timber. Besides biological eruptions, those from volcanoes, allied with other geological phenomena, pushed the climate this way and that, and the sporadic arrival of asteroids served as a dependable killjoy for Earth's captives. Evidence of these processes provides a feeble shelter for people who side with my Texan relation and see climate change – if they are willing to accept that the planet is getting toasty at all – as a non-human phenomenon for which we bear no responsibility.

Humans and other bipedal apes have pursued our distinctively destructive path for a sliver of the total biotime in this corner of the galaxy. This most recent reshaping of nature began 3.3 million years ago, when an australopithecine made stone tools to butcher animal carcasses on the shores of the Jade Sea, or Lake Turkana, in Kenya. Weapons came later, with the use of stone-tipped thrusting spears by another hominid in South Africa 500,000 years ago, and the development of the bow and arrow by early humans 71,000 years ago.[10] Projectile weapons, like the bow and arrow, allowed us to kill large animals without being excessively brave. Through a combination of these weapons, coupled with traps and fire, humans saw to the extinction of woolly mammoths, mastodons, sabre-toothed cats and ground sloths as the ice sheets receded and we pursued the animals to their last redoubts. A South American armadillo-like animal called *Glyptodon* was another victim of the genocide. This slow-moving vegetarian was as big as a Volkswagen Beetle and served as an easy target for hunters who ate its meat and crawled into its enormous shells for shelter.

For many years, biologists argued that climate change was the most important factor in these extinctions, but more and more evidence points to the correspondence between the arrival of humans and the disappearance of large mammals.[11] The case was pretty obvious for the spectacular bird life of islands, with a giant turkey called *Sylviornis* disappearing from New Caledonia soon after the prehistoric Lapita people arrived in their canoes 3,500 years ago, and the elimination of numerous species of flightless moa when the Maori reached New Zealand around AD 1300.[12] Extinction has been reworking nature from its beginnings, but no animal has come close to having the impact that humans have had. With remarkable speed, our evolution walloped life with the power of the asteroid that obliterated the dinosaurs. The average size of mammals increased steadily throughout the Cenozoic Era that followed the crash of the Chicxulub asteroid in the Gulf of Mexico 65 million years ago. Then, around 100,000 years ago, the big animals began to disappear. The extinctions accelerated 50,000 years ago and the total mass of wild mammals has now plunged to a sixth of its pre-human maximum. According to some models, the domestic cow is on track to become the largest remaining mammal.[13]

Scepticism surrounding these doom-laden predictions about the precarious nature of nature is understandable. It takes imagination to escape from the influence of the diminishing expectations of each generation. Nobody has seen a live moa since the fourteenth century and so their absence does not upset New Zealanders today. The last passenger pigeon, named Martha, died in my local zoo in 1914, and the most recent sky-darkening mass migrations of these birds took flight in the nineteenth century. We cannot miss something that has never existed for us. We read about extinction as an approaching horror and ecosystem damage as a work in progress rather than a done deal. But the destruction is unabated. Despite the publicity given to deforestation, tropical woodlands continue to disappear at an annual rate of 2.7 million hectares in Brazil, 1.3 million hectares in Indonesia and 0.6 million hectares in the Democratic Republic of

Congo.[14] Turning to the direct effects of climate change, one-third of the world's coral reefs were damaged by high water temperatures in 2016. More than 90 per cent of Australia's Great Barrier Reef was affected by the process called bleaching, which happens when the dinoflagellate algae abandon their animal partners in the exquisite coral symbiosis.[15] When reefs recover from bleaching, the original animals are replaced by sluggish coral species that support impoverished communities of marine life. This is not a normal phenomenon.

Moving to a more spectacular ecosystem, namely my Ohio garden, we have crafted an Eden amid the trees and flowering shrubs on our triangular plot. The shadiest corners are ferned and bedded with soft moss that teems with amoebae and water bears. Moles sift the soil, fish swim in a pond and a quartet of chickens flop dazed in afternoon dust baths. We have tended this suburban oasis for more than twenty years without a spritz of pesticide, but its biology is changing fast. Many of the glorious insects that were here in the early summers have not returned for a decade. Hummingbird hawkmoths and stick insects have disappeared; cabbage whites are the only butterflies now, and nocturnal moths no longer crowd the evening porch lights. Pure anecdote, yes, but personal observations that align perfectly with scientific surveys that demonstrate startling losses of flying insects.[16]

Larger animals are also affected. My occasional wanderings after dark make me certain that the garden receives fewer visits from racoons, possums and skunks. Little brown bats have become so rare that the appearance of a pair of these lovely mammals at sunset is a cause for jubilation. White nose disease may have killed some animals, and the scarcity of insects must be starving those that escape the fungus. The most obvious change has come with the thinning of trees by an invasive beetle, the emerald ash borer, whose larvae have killed every white ash in the region. The news is no better when we leave the suburbs. Streams on the surrounding farmland are clogged with algae and the big spiders that webbed the crop edges are gone. Even meadow mushrooms have become an oddity. All around us, nature is coming apart at the seams.

The International Union for Conservation of Nature and Natural Resources, or IUCN, compiles the Red List of Threatened Species, which ranks species according to their proximity to extinction. Where data is available, species are assigned categories that run from 'Least Concern' to 'Critically Endangered'. There are two categories for extinct species: extinct in the wild (such as the Hawaiian crow) and extinct (passenger pigeons). The IUCN Red List places *Homo sapiens* in the conservation category of 'Least Concern', and offers the following justification: 'Listed as Least Concern as the species is very widely distributed, adaptable, currently increasing, and there are no major threats resulting in an overall population decline.'[17] Really?

In the unlikely event that we develop the means to control warming, the population will continue to climb but the world we will inhabit will be scrubbed of its biological diversity. The big animals will be gone from the wild. We will find ourselves simultaneously crowded by humans and lonely in nature. This is clear from a random sampling from the IUCN list of endangered and critically endangered species: the humphead wrasse is endangered by reef fishing with spears, explosives and cyanide; the common sawfish is critically endangered by hydroelectric dams, pollution and trophy hunters; the eastern long-beaked echidna is disappearing from New Guinea, where its habitat is destroyed by mining companies, and the great hammerhead is endangered by the annual harvest of an estimated 73 million of these graceful fish to supply the Chinese market for shark fin soup. The only way to conserve these species is to exclude their habitats from all human contact.

Baconian science is at the root of the apocalypse. We have been blessed by advances in medicine, agriculture and engineering. Science has done exactly what we asked of it and now we are set for annihilation. If European science had petered out after the discoveries of the seventeenth century, we would be less numerous and Earth would not be warming. We were cautioned in Genesis, as John Milton reframes the story in *Paradise Lost*:

> Of man's first disobedience, and the fruit
> Of that forbidden tree, whose mortal taste
> Brought death into the world, and all our woe
> With loss of Eden . . . (Book I, 1–4)

Unimpressed by God's warning and encouraged by the guileful serpent, Eve made her fateful decision:

> Forth reaching to the fruit, she plucked, she ate:
> Earth felt the wound, and nature from her seat
> Sighing through all her works gave signs of woe,
> That all was lost. (Book IX, 781–4)

Eve was the first experimentalist, a young woman who tested the boundaries of her environment and sought more than an eternity of servitude in a beautiful garden. Milton lived on the cusp of the scientific revolution and could not have appreciated the power of this metaphor in our time. Should John Snow have burned his 1854 map of Soho that matched cholera cases to contaminated wells?

This would have helped to keep the number of Londoners trim. Perhaps we would have forestalled extinction if Louis Pasteur had abandoned his studies on the germ theory. What about the plant pathologists who scorned centuries of superstitions and identified the fungi responsible for cereal diseases? They made it possible to combat the rusts and smuts that wasted crops and allowed modern agriculture to feed us in our billions.

Science is so central to modern civilization that we will not willingly retreat from the continuing exploration and manipulation of nature. Now that the downside of our loss of innocence is evident, we can burn and rave as Dylan Thomas recommended, or consider plans to step aside with some grace. But whatever else happens, we cannot continue to champion the purity of science without recognizing the terrible cost of discovery; 'For this revolt of thine methinks is like/ Another fall of man' (*Henry V*, Act II, Scene 3).

TEN

GRACE

How We Should Leave

In short order, innovations in energy production and transportation, coupled with advances in agriculture and medicine that sustained population growth, have brought us to this warming world. This perilous outcome is the product of Western science and engineering, founded on the principles of Francis Bacon's experimental methods, and will lead to the collapse of civilization and our eventual extinction. How will we react to this distressing conclusion?

Forecasting the end times, it seems probable that people with any sort of pleasurable lifestyle will seek to protect the status quo for as long as possible and do little to reduce carbon emissions. Like the French aristocracy of the eighteenth century, we will perfect an ethos of inattention and double-down on the pastimes that make us happiest. There will be festivals of disavowal as long as the celebrants can stand the heat. Before long, far from the sound stages and youthful jollity, wars will be fought over usable farmland and sources of fresh water, walls and fences will criss-cross the landscape and armies will be deployed to prevent the poor from flowing across borders.

As the temperature rises, the patricians will seek refuge as polar migrants, or set sail on heavily armed ocean liners. Millions more will live in underground cities, anywhere to escape the sun. Dazzling reports of new methods for sopping up the gigatons of carbon dioxide will create ripples of enthusiasm and then fade in the next news cycle. Fisheries and agriculture will collapse, drugs will provide little solace, and everyone will curl up in a foetal position in the end, like the ash-entombed victims at Pompeii, whimpering in the inescapable heat. The likelihood of this outcome increases as the years pass and the smoke rises.

Denial of planetary damage is surprisingly resilient in these early years of this century, although the objections raised against the multi-layered scientific evidence for the mechanism and progression of planetary roasting seem increasingly clownish.[1] This does not mean, of course, that people who accept the facts are on the same page when it comes to the urgency of the phenomenon. A survey published in 2017 showed that majority of corn (maize) growers in the Midwestern United States recognize that the weather is less predictable than it used to be.[2] They have responded by reducing tillage, planting the latest hybrids and implementing other strategies to protect their fields from more frequent droughts and flooding. They have also increased their crop insurance. Yet they remain quite calm, believing that climate change may not have a significant effect on the profitability of their farms and that future challenges will be solved by human ingenuity. This optimism is understandable among working people who have witnessed astonishing technological innovations in agriculture during their lifetimes. There are even indications that high crop yields will be supported by the warmer and wetter weather conditions that are forecast for the middle of America in the short term.[3]

Farmers are having a tougher time elsewhere. Cereal growers in India have seen their livelihoods, and every hope of a cooler and wetter future, evaporate in the heat of successive summers.[4] Suicide rates have risen among these farmers and we have the makings of an emerging mental health pandemic in the developing world. Less dramatic responses to ecological damage have been studied among indigenous Inuit communities in northern Canada and wheat farmers in Australia.[5] Both groups have been subjected to striking alterations in regional climate, which have brought major changes to their ways of life. Surveys report that these populations are experiencing 'ecological grief' associated with the physical changes in their environment, and people feel hopeless about the future.

Even those who have not themselves felt the effects of warming have grave misgivings on behalf of future generations. In the United States, an increasing number of young women express concerns about having children without 'a little more certainty that there would be a reasonable world for them to inherit'.[6] Each avoided baby is one less person who will suffer and one less carbon footprint. Curbing consumerism might help to improve the environmental prospects, but our genes fight against this, just as surely as so many people continue to trust that the meaning of life lies in manufacturing babies. The situation seems helpless and it probably is.

Concluding that we have crossed the Rubicon and that a technical fix is unlikely to cool the planet, author Roy Scranton recommends that we 'learn to die not as individuals, but as a civilization'.[7] Following similar logic, Canadian physicians Alejandro Jadad and Murray Enkin penned a provocative editorial in the *European Journal of Palliative Care* in 2017, suggesting that we extend the practices of hospice care to human civilization as a whole.[8] The palliative measures include making the international investments necessary to eradicate hunger and shelter the homeless, and committing to a new era of frugality. They argue that conflicts inflamed by diminishing natural resources could be controlled by diverting military spending to a global peacekeeping taskforce. These actions seem like a stretch for a civilization that has never been noted for its cooperation under the best circumstances. Nationalism, which is an example of narcissism writ large, is the more common modus operandum for our species, and tribal conflicts grow as environmental stress increases. If we were to push humanity off its imagined peak of evolutionary progress, is it possible that we might get along with one another a little better, even as the lights go out?

To argue for the importance of recasting of *Homo* in this era of warming, it is useful to recap the essential themes of this book. We live on a Goldilocks planet that has nurtured life as it has sailed through billions of laps around the Sun. Animals evolved from microbes that resembled sperm cells that wriggled in the sea; great apes, or hominids, were born 15 to 20 million years ago; apes like us, called hominins, arose in Africa more recently, and modern humans with fine-boned skeletons have been prancing around for less than 100,000 years. Plants assemble their tissues from carbon dioxide and the power of sunbeams, and we are energized by eating them and the flesh of animals that graze on fruits and vegetables. The digestive system releases small molecules from our food and these are propelled around the body in blood vessels to sustain every cell. The architecture and operation of the body is detailed in a cluttered instruction manual written in 20,000 genes spotted along 2 m of DNA. Construction takes nine months and includes wiring a big brain that endows the owner with a sense of self and the illusion of free will. Ageing of the body is unfaltering; after a few decades, the animal stops working and is decomposed.

The combination of physical dexterity and brain power has allowed humans to manipulate the environment to suit their needs. No other species has this conscious capability. Hands are crucial:

highly intelligent animals with fins and flippers have no capacity to reconstruct their surroundings. In little time at all, advances in science and engineering have supported the rapid expansion of the human population and made modern life luxurious by burning fossil fuels. The attendant transformation of the atmosphere has caused the surface of Earth to warm.

Something very similar may have played out all over the cosmos. If, as it seems probable, life has evolved on other planets, maybe some extraterrestrials have developed levels of technological sophistication that match or exceed our own industriousness. Why then, asked Enrico Fermi, are things so quiet: 'Where is everybody?' Fermi posed this question, or something close to it, during lunch at the Los Alamos National Laboratory in New Mexico in 1950.[9] Edward Teller was sitting at the same table. The irony in this story is stupendous. Fermi was 'the father of the atomic bomb', and Teller went on to become 'the father of the hydrogen bomb'.[10] If his comedic timing had been better, Fermi should have followed his question by looking at Teller, widening his eyes, and saying: 'Ah, yes, of course!'

The *silentium universi*, or Great Silence, has many explanations, and physicists wrestle with the Drake Equation to estimate the likelihood of a contact.[11] The variables in this calculation include things like the rate of star formation (R^*), and L, the length of time over which extraterrestrial civilizations might produce detectable signals. The development of nuclear weaponry may be a popular way to limit L, but I bet that suicide by incinerating fossils is a more commonplace ending for extraterrestrials.

I imagine that children in alien classrooms learn about the universal rule, which states that any life form that develops the technology to extinguish itself will do so in short order. There are steps in this process, comparable to the stages in the development of cancer, from Stage Zero, when the disease is in one place, to Stage Four, where the cancerous cells have spread to other organs. As Christopher Hitchens, whom we met in Chapter Seven, wrote during his malady, 'the thing about Stage Four is that there is no such thing as Stage Five'. As a species, we have hovered around Stage Four for more than 100,000 years. 'How long do the humans have left on Earth?', asks the schoolteacher on Planet Zeta, and noodly appendages are raised with enthusiasm across the classroom.[12]

Each generation has played its part in diminishing the returns for succeeding generations. It would be absurd for me to admonish my father for driving an Alfa Romeo to work in the 1970s rather than saddling up a mule. Now that the atmospheric consequence of driving is apparent, we could make efforts to car pool, but this runs counter to the capitalist desire for personal control. Thinking about curbing carbon emissions is also discouraged when we consider that the damage is done already and that any benefits from stopping now will not be felt for decades.[13] One reaction among online commentators who understand that we would keep getting warmer even if we were able to halt all carbon emissions now is that it may be best to stay the course, enjoy ourselves 'before the whole shithouse goes up in flames', as Jim Morrison said – destroy humanity and allow the planet to reboot in our absence.[14]

The rest of nature will celebrate our departure. If extraterrestrials had trained their microphones on Earth they would have detected a rise in the exclamations of animal life in recent millennia, building to a crescendo of moans and grunts from animals subjected to ritualized torture in stadia, bull rings and bear pits, augmented by the modern vivisection of rodents, cats and primates – terrified animals taped and manacled to chairs and probed with instruments that would have taxed the pornographic inventiveness of the Catholic inquisitors. The philosopher Schopenhauer said: 'Unless *suffering* is the direct and immediate object of life, our existence must entirely fail of its aim.'[15] Today's justifications for these horrors include the economic burden of treating animals more kindly and the medical necessity of experimentation. We rest, as always, on staggering hubris. It is always about us.

This lack of empathy with other animals is at odds with the idea that we have an instinctive love of nature, which has been described as 'biophilia'. Biophilia was popularized by the Harvard biologist E. O. Wilson, who suggested that we maintained feelings of empathy from our prehistoric engagement with wildlife on the grasslands of Africa.[16] Evidence for this behaviour is non-existent, however, and the concept does not make any evolutionary sense.[17] Humans are as kind towards the natural world as our destruction of it suggests. Every child that enjoys turning over rocks in a stream has a friend who recoils in horror at the sight of a frog or claw-waving crayfish.[18] If there is anything that is instinctive, it is the propensity to chase and to kill. Educational programmes on natural history can work wonders in reprogramming the behaviour of children who might otherwise become lifelong biophobes, but many more will resist the charms of bird-watching as long as other distractions are available.

Producers of wildlife documentaries have encouraged decades of wishful thinking that their programmes could tap into some mysterious reverence for nature and help to save the planet. We were thrilled to experience the splendour of a tropical rainforest on television and saddened by the brief segment at the end of the programme showing logging trucks roaring along dusty roads. The better zoos approach their visitors in a similar fashion, displaying animals for their entertainment value and listing their endangered status on an explanatory label attached to the fence or glass.

Children shriek at the gorillas, enjoy an ice cream and are driven home. Evidence that zoos stimulate a lasting passion for animal conservation is weak.[19]

Conservation biologists may empathize with other forms of life, but they do almost as much damage to the planet as their biophobic neighbours. A charitable contribution here and there is not going to change the outcome. Solar panels and electric cars are funeral decorations for Earth. One of the difficulties is that we inflict the greatest wounds passively, through the simple acts of getting on with modern life. John Lennon said, 'Life is what happens to you while you're busy making other plans.' So does climate change. As they wait for Godot, Estragon says, 'I can't go on like this.' Vladimir replies: 'That's what you think.'[20]

Human selfishness has put us in this unenviable situation of presiding over the collapse of the biosphere. To be present at this historical juncture is to be in a position of unwelcome exclusivity, like the predicament of the Romans who happened to be living close to Mount Vesuvius in AD 79. There was no silver lining to witnessing the Great Plague in the fourteenth century either. Like me, the pus-swollen victims of the plague believed that they were confronting the end of civilization, as well their own deaths. At least we will not have to feel worried about eternal damnation. And in the face of this astonishing situation, perhaps we will finally overcome our persistent narcissism. Celebrity or peasant, nothing is going to save you and nobody will be here to care about your legacy in the future. You may have sold millions of books or albums, filled stadiums with fans of your athletic prowess, but nobody will be around soon to give a fig.

Grace, by which I mean thankfulness for the experience of consciousness and our fleeting participation in nature, seems the most soothing mental exercise. 'To go nobly lends a man some grace,' says the leader of the Old Men of Argos in Aeschylus' *Agamemnon*.[21] He is speaking to Cassandra, who has realized that she will be murdered. Expressions of grace have always served this function in confronting one's personal demise and their value need not diminish when we accept that civilization is approaching its end date. The focus on the grand carnival of nature is what's different now, looking squarely at the thing we have spoiled. We gain some personal sense of deliverance in this admission of fault, even though our victim is the whole of creation and we are among the casualties.

In *Paradise Lost*, Eve suggests a pact with Adam as she begins to comprehend death as the penalty for her fall from grace: 'Why stand we longer shivering under fears[?]' (book x, 1003). She is terrified for herself and on behalf of their future offspring, and reasons that suicide will bring an end to the punishment, 'So death be deceived his glut, and with us two/ Be forced to satisfy his rav'nous maw' (990–91). In the end, the first couple opt to accept the punishment as it was imposed and to pursue parenthood in the meantime. Adam and Eve conformed to their programming. We are following suit, unable or unwilling to change course. The best that any of us can do until the sky falls is to be kinder to each other and humane towards the rest of nature as it suffers with us on this watery globe. And who knows, if we are nicer, maybe things will keep running for longer than we expect.

REFERENCES

PREFACE

1 Yuval Noah Harari, *Homo Deus: A Brief History of Tomorrow* (London, 2016).
2 The Latin translation of my species description is by classical scholar Michael Klabunde, Professor Emeritus, Mount St Joseph University.
3 P. G. Wodehouse, *My Man Jeeves* (London, 1919), Chapter Two.

ONE

GLOBE

How Life Lends Itself to Earth

1 Skin is a useful metaphor. Beneath the surface of the largest humans, the tissues run as deep, speaking relatively, as the interior of the planet. The average thickness of the biosphere is 5 km (3 mi.) and the radius of Earth is 6,371 km (3,959 mi.). This means that life exists within the outermost 0.3 per cent of the planet. Human skin ranges from 0.5 to 4.0 mm (0.02 to 0.16 in.) in thickness. Someone with a waist circumference of 200 cm (79 in.) would have an internal radius of 32 cm (13 in.). Wrapped in a 1-mm-thick skin, this individual would provide a live geometric model of the planet. If we are satisfied with

order-of-magnitude comparisons, humans with more modest waist sizes have roughly the same ratio of skin-to-tissue thickness as Earth's biosphere to the planet's innards.

2 Carbon forms when medium-sized stars like our sun collapse, as well as from the explosion of larger stars that become supernovae. The generation of the heaviest elements, like gold and uranium, may require the services of neutron stars, whose collisions release more energy than supernovae and disturb the fabric of space with gravitational waves.

3 The Roman senator Boethius thought it best to be meditative during imprisonment. He was placed under house arrest in northern Italy in AD 523 for treasonous behaviour and was clubbed to death by his Ostrogoth captors. During his confinement he found solace in Greek philosophy and wrote *The Consolation of Philosophy*, trans. Patrick G. Walsh (Oxford, 1999).

4 See David Benatar, *Better Never to Have Been: The Harm of Coming into Existence* (Oxford, 2006).

5 The Milky Way is a middle-sized spiral galaxy and it takes us about 230 million years to make it all the way round at a speed exceeding 800,000 km (500,000 mi.) per hour.

6 See Roger Highfield, 'Colonies in Space May Be the Only Hope, says Hawking', www.telegraph.co.uk, 16 October 2001.

7 The phrase comes from Basil Willey (1897–1978), professor of English literature at Cambridge University.

8 Planck time is regarded by physicists as the shortest interval of time that has any meaning. It is the time taken for light to travel over a distance of one Planck length in a vacuum. One Plank length is 1.62×10^{-35} metres, and Planck time, or one Planck second, lasts for 5.39×10^{-44} seconds. The units of Planck time

and Planck length were proposed by German physicist Max Planck (1858–1947).

9 Intact areas of the ozone layer are just 3 mm in thickness. The proposition that this is 'the best of all possible worlds', by German philosopher Gottfried Leibniz, was satirized by Voltaire in his magnificent novel *Candide, ou l'Optimisme* (Paris, 1759).

10 Thomas Hobbes, *Leviathan*, ed. Noel Malcolm (Oxford, 2012), vol. II, p. 135.

TWO

GENESIS

How We Arrived

1 Ovid, *Metamorphoses*, trans. Arthur Golding (London, 2002), Book I, lines 101–2.

2 Joseph Conrad, *Heart of Darkness* (London, 1983), p. 66.

3 'Choano' is taken from the Greek word *khoane*, for funnel, because the collar is shaped like a funnel; 'flagellate' comes from the Latin *flagellum*, for whip.

4 The function of junction proteins in collar flagellates is not settled, but bacterial recognition and capture is a compelling answer. See Scott A. Nichols et al., 'Origin of Metazoan Cadherin Diversity and the Antiquity of the Classical Cadherin/β-Catenin Complex', *Proceedings of the National Academy of Sciences*, CIX (2012), pp. 13046–51.

5 Making the shaft and handle would come last because the proto-umbrella would still work as a shield from the rain without them. In the same way, the evolution of an eye lens followed the development of pigments that react to light. With a suitable chemical messaging system, a simple spot of pigment could

inform the creature of sunrise and prove beneficial without a lens.

6 Roberto Feuda et al., 'Improved Modeling of Compositional Heterogeneity Supports Sponges as Sister to All Other Animals', *Current Biology*, XXVII (2018), pp. 3864–70.

7 Members of this super grouping of kingdoms are called opisthokonts. This easily forgettable name refers to things with tails that point backwards, from the Greek words *opisthios*, meaning rear, and *kontos*, for pole, like the one used by a gondolier. These tails are flagella. We are opisthokonts and so are fragrant truffles and the sows that unearth them. Truffles and gilled mushrooms do not produce motile cells, but aquatic fungi produce swimming cells with the trademark flagellum that indicates the shared history of the animals and fungi.

8 Some years ago, it was a pleasure, of sorts, to be given the opportunity to critique an institution called the Creation Museum on a live broadcast by a National Public Radio station in Ohio. This disgraceful 'museum' – it is a church, really – is located in Kentucky and preaches the fantastical idea that Earth is 6,000 years old and was formed in harmony with the biblical account. Rejecting the possibility of metaphor on the part of the Hebrew authors of the Book of Genesis, the Creation Museum sticks with a literal reading of the feverish six-day carnival that produced the planet and its biology. I made an eloquent and impassioned case against everything uttered by the museum's founder, an Australian gentleman called Ken Ham, who remained steadfast in his silliness and seemed particularly upset by my description of human beings as a type of ape. Looking back, I wish I had followed up with the scientific case for our

shared ancestry with bath sponges and mushrooms. Sparks would have flown from his head.

9 T. D. Kenny and P. L. Beales, *Ciliopathies: A Reference for Clinicians* (Oxford, 2014).

10 J.A.R. Tibbles and M. M. Cohen, 'The Proteus Syndrome: The Elephant Man Diagnosed', *British Medical Journal*, CCXCIII (1986), pp. 683–5; Marjorie J. Lindhurst et al., 'A Mosaic Activating Mutation *AKT1* Associated with the Proteus Syndrome', *New England Journal of Medicine*, CCCLXV (2011), pp. 611–19.

11 Merrick quoted these lines at the end of some of his letters. They are adapted from a poem by Isaac Watts titled 'False Greatness', published in *Horae Lyricae: Poems, Chiefly of the Lyric Kind, in Two Books* (London, 1706), pp. 107–8.

THREE

GUTS

How Our Bodies Work

1 *Lucian*, vol. VI, trans. K. Kilburn, Loeb Classical Library (Cambridge, MA, 1959), p. 177. Herodotus claims that Philippides ran from Athens to Sparta before the battle; see Herodotus, *The Landmark Herodotus: The Histories*, trans. Andrea L. Purvis (New York, 2007), book VI, Chapter 106, p. 469. This is commemorated in an annual ultramarathon called the Spartathlon. The current record of twenty hours and 25 minutes is held by Yiannis Kouros, a Greek runner known as the 'Running God', who won the race in 1984.

2 As I wrote in *Mr Bloomfield's Orchard: The Mysterious World of Mushrooms, Molds, and Mycologists* (New York, 2002), p. 21: 'If

fungi can rot bone in a patient's leg, feast on someone's brain, or devour a child's face, is the naked ape king of the jungle, or the king's dinner?' Thomas Hobbes expressed the same idea in *The Questions Concerning Liberty, Necessity, and Chance* (London, 1656), p. 141: 'When a [Lion] eats a Man, and a Man eats an Oxe, why is the Oxe more made for the Man, than the Man for the Lion?'

3 Mashed potatoes scored three times higher than white bread on the satiety index of common foods. The satiety index is a measure of the ability of different food items to satisfy appetite: Susanna H. Holt et al., 'A Satiety Index of Common Foods', *European Journal of Clinical Nutrition*, XLIX (1995), pp. 675–90.

4 The reactions that occur during the synthesis of chlorophyll and haemoglobin molecules (both are examples of porphyrins) are among the greatest feats of natural enzyme chemistry. These reactions would take 2.3 billion years to occur on their own, compared with fractions of a second with the assistance of enzymes. See C. A. Lewis and R. Wolfenden, 'Uroporphyrinogen Decarboxylation as a Benchmark for the Catalytic Proficiency of Enzymes', *Proceedings of the National Academy of Sciences*, CV (2008), pp. 17328–33.

5 H. F. Helander and L. Fändriks, 'Surface Area of the Digestive Tract – Revisited', *Scandinavian Journal of Gastroenterology*, XLIX (2014), pp. 681–9.

6 Following a school biology class in which we looked at a diagram of an evolutionary tree, the same boy suggested to me that 'evolution did not make sense', because there are plenty of amoebas around today. 'When', he asked, 'are they going to evolve into humans?' He is probably running a Fortune 500 company today.

7 Eva Bianconi et al., 'An Estimation of the Number of Cells

in the Human Body', *Annals of Human Biology*, XL (2013), pp. 463–71, estimated that an adult human is built from 3.72×10^{13} cells.

8 An average heart rate of 70 beats per minute corresponds to 100,800 beats per day. The estimated length of the circulatory system comes from Benjamin W. Zweifach, 'The Microcirculation of the Blood', *Scientific American*, CC (1959), pp. 54–60.

9 The removal of electrons is called oxidation. An oxidation reaction is balanced by a reduction reaction in which electrons are accepted by another substance. Together, the reactions of oxidation and reduction constitute redox chemistry. Oxygen is one of many compounds that serve as oxidizing agents, meaning that they accept electrons from other substances. Oxygen is only involved in the final reaction of sugar metabolism, when it accepts electrons and combines with protons (= charged hydrogen atoms, symbolized as H^+) to form water.

10 Dominique Chrétian et al., 'Mitochondria are Physiologically Maintained at Close to 50°C', *PLOS Biology*, XVI/1 (2018), e2003992.

11 Heidi S. Mortensen et al., 'Quantitative Relationships in Delphinid Neocortex', *Frontiers in Neuroanatomy*, VIII (2014), DOI: 10.3389/fnana.2014.00132.

12 Irene M. Pepperberg, 'Further Evidence for Addition and Numerical Competence by a Grey Parrot (*Psittacus erithacus*)', *Animal Cognition*, XV (2012), pp. 711–17.

13 D. M. Bramble and D. E. Lieberman, 'Endurance Running and the Evolution of *Homo*', *Nature*, CDXXXII (2004), pp. 345–52.

14 Michael Roggenbuck et al., 'The Microbiome of New World Vultures', *Nature Communications*, V/5498 (2014), DOI: 10.1038/ncomms6498.

15 Karen Hardy et al., 'The Importance of Dietary Carbohydrate in Human Evolution', *Quarterly Review of Biology*, XC (2015), pp. 251–68.

16 Herman Pontzer et al., 'Primate Energy Expenditure and Life History', *Proceedings of the National Academy of Sciences*, CXI (2014), pp. 1433–7.

17 Current data available at www.worlddata.info and www.indexmundi.com.

18 The philosophical concept of plenism, or *horror vacui*, originated with Aristotle. The phrase was restated as *natura abhorret vacuum* by François Rabelais in *The Life of Gargantua and Pantagruel*, a series of five novels published between 1532 and 1564.

19 Viruses encode even more information in their DNA and RNA molecules than any species made from cells. Most biologists treat viruses separately from cellular organisms because it is questionable whether they count as living things at all. See Nicholas P. Money, *Microbiology: A Very Short Introduction* (Oxford, 2014), p. 18.

20 Shoukat Afshar-Sterle et al., 'Fas Ligand-mediated Immune Surveillance by T Cells is Essential for the Control of Spontaneous B Cell Lymphomas', *Nature Medicine*, XX (2014), pp. 283–90.

21 Hans Zinsser, *Rats, Lice and History* (Boston, MA, 1935), p. 185. Zinsser supports the general thesis of my book in his 1935 classic: 'Man and rats are merely, so far, the most successful animals of prey. They are utterly destructive of other forms of life. Neither of them is of the slightest use to any other species of living things.'

22 This description is paraphrased from Nicholas P. Money, *The Rise of Yeast: How the Sugar Fungus Shaped Civilization*

(Oxford, 2018), p. 172. More than half of the human body's weight is carried as water, one-fifth is protein, another fifth is fat and the balance is made up by the minerals that form the skeleton. This information comes from the chemical analysis of cadavers: Harold H. Mitchell et al., 'The Chemical Composition of the Adult Human Body and its Bearing on the Biochemistry of Growth', *Journal of Biological Chemistry*, CLVIII (1945), pp. 625–37; Steven B. Heymsfield et al., eds, *Human Body Composition*, 2nd edn (Champaign, IL, 2005).

FOUR

GENES
How We Are Programmed

1 DNA double helix thickness is 2 nanometres (nm) or 2×10^{-9} m; the total length of DNA per human cell is 2 m; the ratio of thickness to length is $1:10^9$. Pencil thickness is 8 mm, so the corresponding length is 8×10^9 mm, or 8,000 km. To match the compactness of a nucleus, this 8,000-km-long pencil would have to be folded into a hot tub.
2 The anatomist William Harvey (1578–1657) adopted this view in the 1650s, and Francesco Redi (1626–1697) offered experimental proof, at least for insects, with his detailed descriptions of larvae hatching from eggs.
3 Mycoplasmas are among the smallest bacteria with a cell diameter of between 0.2 µm and 0.3 µm (200–300 nm). The micrometre unit, symbolized 'µm', represents one-millionth of 1 m. Microbiologists have identified even smaller bacteria from groundwater samples in Colorado, but very little is known about

their biology. The infectious particles of viruses are considerably smaller than bacteria, but these are not cells. Blue whales are the largest animals that have ever existed, but they may be outweighed by the collective mass of microscopic filaments produced by colonies of fungi that form mushrooms. See Nicholas P. Money, *Mushrooms: A Natural and Cultural History* (London, 2017).

4 The adjective 'extraordinary' has been rendered meaningless by English commentators on BBC radio and television. Scientists and historians fling the term around to the point that whatever they are describing cannot, by definition, be extraordinary at all. If this species of seashell is truly extraordinary, others cast on the beach cannot be so.

5 The microscopist Antonie van Leeuwenhoek (1632–1723) provided a serious distraction from progress in understanding human reproduction with his belief that the sperm carried miniature unborn children, and that mothers served as mere incubators. Leeuwenhoek's contemporary, Nicolaas Hartsoeker, developed this idea with conceptual drawings of the tiny man, or homunculus, that lived in the head of sperm cells. Leeuwenhoek and Hartsoeker did not claim to have seen the homunculus themselves. See Kenneth A. Hill, 'Hartsoeker's Homunculus: A Corrective Note', *Journal of the History of the Behavioral Sciences*, XXI (1985), pp. 178–9.

6 Larger genomes may exist, but the canopy plant has the biggest that has been verified by sequencing the entire thing.

7 Jean-Michel Claverie, 'What If There Are Only 30,000 Human Genes?', *Science*, CCXCI (2001), pp. 1255–7.

8 A. F. Palazzo and T. R. Gregory, 'The Case for Junk DNA', *PLOS Genetics*, X/5 (2014), e1004351.

9 The 1000 Genomes Project Consortium, 'A Global Reference for Human Genetic Variation', *Nature*, DXXVI (2015), pp. 68–74.

10 Ning Yu et al., 'Larger Genetic Differences within Africans than between Africans and Eurasians', *Genetics*, CLXI (2002), pp. 269–74; L. B. Jorde and S. P. Wooding, 'Genetic Variation, Classification and "Race"', *Nature Genetics Supplement*, XXXVI (2004), pp. S28–33.

11 Carl C. Bell, 'Racism: A Symptom of the Narcissistic Personality Disorder', *Journal of the National Medical Association*, LXXII (1908), pp. 661–5.

FIVE

GESTATION

How We Are Born

1 Jamie A. Davies, *Life Unfolding: How the Human Body Creates Itself* (Oxford, 2014), provides a nice introduction to human embryology.

2 *Volvox* is a beautiful green alga with spherical colonies whose surface is occupied by hundreds or thousands of cells. Each cell is equipped with a pair of cilia and their activity is coordinated to propel the colony through water. The colonies revolve slowly as they swim and their motion has been neatly described as planetary. *Volvox* reproduces by forming miniature colonies inside the mother sphere. The cells of a new colony and their cilia face towards the inside of their tiny globes. At maturity, each of the mini colonies becomes depressed on one side, and then turns inside out, displaying the cilia on the outside where they function in swimming. This eversion process resembles gastrulation and is recognized as a model in embryological research. See

R. Schmitt and M. Sumper, 'Developmental Biology: How to Turn Inside Out', *Nature*, CDXXIV (2003), pp. 499–500.

3 Structures called the pit and node form at one end of the primitive streak. The node is the structure mentioned in Chapter Two, in which fluid movements driven by cilia are involved in the foundation of the left–right axis.

4 Janet Rossant, 'Human Embryology: Implantation Barrier Overcome', *Nature*, DXXXIII (2016), pp. 182–3.

5 I. Hyun, A. Wilkerson and J. Johnston, 'Embryology Policy: Revisit the 14-day Rule', *Nature*, DXXXIII (2016), pp. 169–71.

6 B. Prud'homme and N. Gompel, 'Evolutionary Biology: Genomic Hourglass', *Nature*, CDLXVIII (2010), pp. 768–9.

7 Robert L. Stevenson, *The Strange Case of Dr Jekyll and Mr Hyde* [1886] (New York, 1980), p. 122. The brilliant 1927 novel *Steppenwolf*, by Herman Hesse, treads a path of comparable uncertainty with the concept of human evolutionary progress.

8 Jean-Baptiste De Panafieu and Patrick Gries, *Evolution*, trans. Linda Asher (New York, 2011). The gorgeous photographs by Patrick Gries of animal skeletons on a flat black background illustrate both the diversity and the unity of animals.

9 Herman Melville, *Moby-Dick; or, The Whale* [1851] (New York, 1992), p. 424.

10 Edmund Spenser, *The Faerie Queene* [1590] (London, 1987), Book I, Canto VI, 1–9.

11 Karl H. Teigen, 'How Good is Luck? The Role of Counterfactual Thinking in the Perception of Lucky and Unlucky Events', *European Journal of Social Psychology*, XXV (1995), pp. 281–302.

12 Morgane Belle et al., 'Tridimensional Visualization and Analysis of Early Human Development', *Cell*, CLXIX (2017), pp. 161–73.

13 David J. Mellor et al., 'The Importance of "Fetal Awareness" for Understanding Pain', *Brain Research Reviews*, XLIX (2005), pp. 455–71.

14 Many writers have wrestled with the unfairness of sanctioning animal suffering while criminalizing human abortion. The arguments are treated with admirable balance in Sherry F. Colb and Michael C. Dorf, *Beating Hearts: Abortion and Animal Rights* (New York, 2016).

SIX
GENIUS
How We Think

1 The soul remains a necessary hypothesis for Christian theologists who contend that neuroscience cannot explain the grandeur of our emotional lives. They ask us, without offering an explanation, to accept that souls deal with the more glorious emotions like love and leave the daily tasks of remembering to go to the post office to the brain. If you take the time to untangle these ideas, it becomes clear that the soul is defended by theologists solely because its existence is required by their faith. Soulless, we have no greater prospects of immortality than an earthworm, which is an idea that has never bothered me in the slightest.

2 African elephant brains weigh 4.5–5 kg (10–11 lb); a sperm whale brain weighs 8 kg (18 lb).

3 The two hemispheres of the human cortex have a combined surface area of 0.24 sq. m (2.6 sq. ft). If the wrinkles were flattened to produce a smooth hemisphere, its diameter would be 0.39 m (15 in.).

4 Michael O'Shea, *The Brain: A Very Short Introduction* (Oxford, 2006).

5 J. Polimeni and J. P. Reiss, 'The First Joke: Exploring the Evolutionary Origins of Humor', *Evolutionary Psychology*, IV (2006), pp. 347–66; R. Rygula, H. Pluta and P. Popik, 'Laughing Rats Are Optimistic', *PLOS ONE*, VII/12 (2012), e51959.

6 Gillian M. Morriss-Kay, 'The Evolution of Human Artistic Creativity', *Journal of Anatomy*, CCXVI (2010), pp. 158–76. English white male bias notwithstanding, but Turner and Milton sprang first to my mind and it would be disingenuous to have substituted a painter and a poet from another culture.

7 'Cartesian' refers to the Latinized name of Descartes – *Cartesius*.

8 A. B. Barron and C. Klein, 'What Insects Can Tell Us about Consciousness', *Proceedings of the National Academy of Sciences*, CXIII (2016), pp. 4900–908; C. J. Perry, A. B. Barron and L. Chittka, 'The Frontiers of Insect Cognition', *Current Opinion in Behavioral Sciences*, XVI (2017), pp. 111–18.

9 Nicholas J. Strausfeld et al., 'Evolution, Diversity, and Interpretations of Arthropod Mushroom Bodies', *Learning and Memory*, V (1998), pp. 11–37.

10 Kevin Healy et al., 'Metabolic Rate and Body Size Are Linked with Perception of Temporal Information', *Animal Behavior*, LXXXVI (2013), pp. 685–96; Rowland C. Miall, 'The Flicker Fusion Frequencies of Six Laboratory Insects and the Response of the Compound Eye to Mains Fluorescent "Ripple"', *Physiological Entomology*, III (1978), pp. 99–106.

11 Arthur Schopenhauer, *Essay on the Freedom of the Will*, trans. Konstantin Kolenda (Mineola, NY, 2005), p. 24. Emphasis in original.

12 Andrew Gordus et al., 'Feedback from Network States Generates Variability in a Probabilistic Olfactory Circuit', *Cell*, CLXI (2015), pp. 215–27.

13 T. Brunet and D. Arendt, 'From Damage Response to Action Potentials: Early Evolution of Neural and Contractile Modules in Stem Eukaryotes', *Philosophical Transactions of the Royal Society B*, CCCLXXI: 20150043 (2015), DOI: 10.1098/rstb.2015.0043; P. Calvo and F. Baluška, 'Conditions for Minimal Intelligence across Eukaryote: A Cognitive Science Perspective', *Frontiers in Psychology*, VI (2015), DOI: 10.3389/fpsyg.2015.01329.

14 Slime mould study: R. P. Boisseau, D. Vogel and A. Dussutour, 'Habituation in Non-neural Organisms: Evidence From Slime Moulds', *Proceedings of the Royal Society B*, CCLXXXIII (2016), DOI: 10.1098/rspb.2016.0446; an alga with an eye: T. A. Richards and S. L. Gomes, 'Protistology: How to Build a Microbial Eye', *Nature*, DXXIII (2015), pp. 166–7; mushroom colony calculation based on data drawn from G. W. Griffith and K. Roderick, in *Ecology of Saprotrophic Basidiomycetes*, ed. L. Boddy, J. C. Frankland and P. van West (London, 2008), pp. 277–99, and Nicholas P. Money, *Mushroom* (New York, 2011). Single cells of algae and fungi display their sensitivity to researchers who penetrate their cells with glass needles, or micropipettes, to make a variety of measurements. The cells respond immediately by sealing off the tip of the needle with a shower of cytoplasmic globs in a kind of immune reaction to attack.

15 Fossils from fifteen individuals of *Homo naledi* were found in a cave system in 2013 and 2014; Lee R. Berger et al., '*Homo naledi*, a New Species of the Genus *Homo* from the Dinaledi Chamber,

South Africa', *eLife*, IV (2015), e09560; Paul H.G.M. Dirks et al., 'The Age of *Homo naledi* and Associated Sediments in the Rising Star Cave, South Africa', *eLife*, VI (2017), e24231.

16 Cosimo Posth et al., 'Deeply Divergent Archaic Mitochondrial Genome Provides Lower Time Boundary for African Gene Flow into Neanderthals', *Nature Communications*, VIII (2017), DOI: 10.1038/ncomms16046.

17 Thomas Hobbes, *De Cive: The English Version* [1651] (Oxford, 1983), p. 34.

18 George R. Pitman, 'The Evolution of Human Warfare', *Philosophy of the Social Sciences*, XLI (2011), pp. 352–79; José M. Gómez et al., 'The Phylogenetic Roots of Human Lethal Violence', *Nature*, DXXXVIII (2016), pp. 233–7.

19 W. Gilpin, M. W. Feldman and K. Aoki, 'An Ecocultural Model Predicts Neanderthal Extinction through Competition with Modern Humans', *Proceedings of the National Academy of Sciences*, CXIII (2016), pp. 2134–9; Thomas Sutikna et al., 'Revised Stratigraphy and Chronology for *Homo floresiensis* at Liang Bua in Indonesia', *Nature*, DXXXII (2016), pp. 366–9.

20 Marina Warner described Joan of Arc as a dancer in a metaphorical sense in *Joan of Arc: The Image of Female Heroism* (Berkeley, CA, 1981). We know so little about the teenage 'Maid of Orléans' that picturing her as a dancer is pure fantasy.

SEVEN

GRAVES

How We Die

1 Christopher Isherwood, *A Single Man* (New York, 1964), p. 186.

2 Beginning with a founding population of three women and

three men, and assuming that each woman had three babies before age twenty, and three afterwards, this uberous tribe would swell to 1 trillion in five centuries. If parents died at any age after producing their children, then it would take 490 years to hit a trillion. If nobody ever died, the 1 trillion mark would be hit only three years sooner. Ergo, death after parenting has little impact on population growth.

3 Christopher Marlowe, *The Tragical History of Dr Faustus* [1592] (London, 1993), A-text, Scene 15, p. 72.

4 Michael R. Rose, *Evolutionary Biology of Aging* (New York, 1991).

5 Drawing on earlier work by Ronald A. Fisher and John B. S. Haldane, Peter Medawar wrote: 'If a genetical disaster . . . happens late enough in individual life, its consequences may be completely unimportant.' This means that 'the force of natural selection weakens with increasing age'; Peter B. Medawar, *An Unsolved Problem in Biology* (London, 1952), p. 18. Medawar went on to suggest that any post-reproductive period of life represents 'a dustbin for the effects of deleterious genes' (p. 23).

6 'Foist' is an appropriate verb because we do not have the opportunity to ask the next generation for permission.

7 There are no 'death genes' that promote ageing and facilitate the death of elderly organisms, but death after mating appears to be programmed in the DNA of cannibalistic fishing spiders. After the male spider transfers sperm into the female, he remains attached to his mate, curls up and dies. This developmental process is advantaged because it avoids the energy expenditure involved in attacking a mate that is reluctant to be eaten. The resulting economy increases the likelihood that female spiders will produce a large number of healthy babies, and the spiderlings honour their gallant fathers by conveying their

genes into the future. See S. K. Schwartz, W. E. Wagner and E. A. Hebets, 'Spontaneous Male Death and Monogyny in the Dark Fishing Spider', *Biology Letters*, IX (2013), DOI: 10.1098/RSBL.2013.0113; and by the same authors, 'Males Can Benefit from Sexual Cannibalism Facilitated by Self-sacrifice', *Current Biology*, XXVI (2016), pp. 2794–9.

8 J. W. Shay and W. E. Wright, 'Hayflick, His Limit, and Cellular Ageing', *Nature Reviews Molecular Cell Biology*, I (2000), pp. 72–6; R. DiLoreto and C. T. Murphy, 'The Cell Biology of Aging', *Molecular Biology of the Cell*, XXVI (2015), pp. 4524–31; Hyeon-Jun Shin et al., 'Etoposide Induced Cytotoxicity Mediated by ROS and ERK in Human Lidney Proximal Tubule Cells', *Scientific Reports*, VI (2016), DOI: 10.1038/srep34064.

9 Leonard Hayflick, 'Entropy Explains Aging, Genetic Determinism Explains Longevity, and Undefined Terminology Explains Misunderstanding Both', *PLOS Genetics*, III/12 (2005), e220, doi.org/10.1371/journal.pgen.0030220.

10 Emily Dickinson, 'Poem 605', in *The Poems of Emily Dickinson*, ed. Ralph W. Franklin (Cambridge, MA, 1998), pp. 601–2.

11 Christopher Hitchens, *Mortality* (New York, 2012), p. 7. Christopher died in 2011, and, like Milton's hero in *Lycidas*, 'hath not left his peer'.

12 S. Jay Olshansky, 'Ageing: Measuring our Narrow Strip of Life', *Nature*, DXXXVIII (2016), pp. 175–6; X. Dong, B. Milholland and J. Vijg, 'Evidence for a Limit to Human Lifespan', *Nature*, DXXXVII (2016), pp. 257–9.

13 M. Depczynski and D. R. Bellwood, 'Shortest Recorded Vertebrate Lifespan Found in a Coral Reef Fish', *Current Biology*, XV (2005), R288–9; Julius Nielsen et al., 'Eye Lens Radiocarbon Reveals Centuries of Longevity in the Greenland

Shark (*Somniosus microcephalus*)', *Science*, CCCLIII (2016), pp. 702–4.

14 M. P. Gardner, D. Gems and M. E. Viney, 'Aging in a Very Short-lived Nematode', *Experimental Gerontology*, XXXIX (2004), pp. 1267–76; Paul G. Butler et al., 'Variability of Marine Climate on the North Icelandic Shelf in a 1,357-year Proxy Archive Based on Growth Increments in the Bivalve *Arctica islandica*', *Palaeogeography, Palaeoclimatology, Palaeoecology*, CCCLXXIII (2013), pp. 141–51.

15 Lucretius, *De Rerum Natura (On the Nature of Things)*, Book III, 972–5, trans. William H. D. Rouse, revd Martin F. Smith, Loeb Classical Library (Cambridge, MA, 1992), pp. 264–5. Lucretius advanced the symmetry argument, positing that the fear of death is irrational: 'Look back also and see how the ages of everlasting time past before we were born have been to us nothing. This therefore is a mirror which nature holds up to us, showing the time to come after we at length shall die.'

16 Thomas M. Bartol et al., 'Nanoconnectomic Upper Bound on the Variability of Synaptic Plasticity', *eLife*, IV (2015), e10778.

17 M. Kaeberlein, C. R. Burtner and B. K. Kennedy, 'Recent Developments in Yeast Aging', *PLOS Genetics*, III/5 (2007), e84.

18 Ferdinando Boero, 'Everlasting Life: The "Immortal" Jellyfish', *The Biologist*, LXIII/3 (2016), pp. 16–19.

19 These lines are sung by Guiderius, eldest son of Cymbeline, at the 'funeral' for Imogen, his sister, who he thinks is a boy, and is not dead, but has been drugged by her stepmother. A similar sentiment was expressed in the song 'Dust in the Wind', by the progressive rock band Kansas, and cried rather beautifully by lead singer Steve Walsh in the late 1970s of my memorable teens.

20 P. Bjerregaard and I. Lynge, 'Suicide – A Challenge in Modern Greenland', *Archives of Suicide Research*, X (2006), pp. 209–20; P. Bjerregaard and C.V.L. Larsen, 'Time Trend by Region of Suicides and Suicidal Thoughts among Greenland Inuit', *International Journal of Circumpolar Health*, LXXIV (2015), DOI: 10.3402/ijch.v74.26053.

21 Lizbeth González-Herrera et al., 'Studies on RNA Integrity and Gene Expression in Human Myocardial Tissue, Pericardial Fluid and Blood, and its Postmortem Stability', *Forensic Science International*, CCXXXII (2013), pp. 218–28; Ismail Can et al., 'Distinctive Thanatomicrobiome Signatures Found in the Blood and Internal Organs of Humans', *Journal of Microbiological Methods*, CVI (2014), pp. 1–7. Note for biologists: after death, tissues in the heart and elsewhere lose their energizing mix of oxygen and glucose from the bloodstream. To survive, they must rely on stored fatty acids and glycogen, and use glycolysis to supply the ATP needed for gene expression.

22 The phrase 'salt of our youth' is from Shakespeare's *The Merry Wives of Windsor*, Act II, Scene 3, voiced by Robert Shallow, a wealthy landowner who claims, 'I have lived fourscore years and upward,' in Act III, Scene 1.

23 Jessica L. Metcalf et al., 'Microbial Community Assembly and Metabolic Function during Mammalian Corpse Decomposition', *Science*, CCCLI (2016), pp. 158–62.

24 Nicholas P. Money, *The Amoeba in the Room: Lives of the Microbes* (Oxford and New York, 2014), pp. 131–52.

25 A. A. Milne, *The House at Pooh Corner* (London, 1928).

EIGHT

GREATNESS

How We Make Things Better

1 D. Castelvecchi and A. Witze, 'Einstein's Gravitational Waves Found at Last', *Nature News* (11 February 2016), DOI: 10.1038/nature.2016.19361.

2 Carl Sagan said this memorable phrase in an American television interview in 1996: 'Science is more than a body of knowledge, it's a way of thinking; a way of sceptically interrogating the universe with a fine understanding of human fallibility.'

3 This quote is taken from the Preface to the *Instauratio Magna*, which was Francis Bacon's uncompleted plan for the reinvention and renewal of natural philosophy. *Novum Organum*, which was published in 1620, was the second part of this larger uncompleted work.

4 Francis Bacon, *Novum Organum* [1620], Book I, LIV (Franklin Center, PA, 1980), p. 234.

5 James D. Watson, *The Annotated and Illustrated Double Helix*, ed. A. Gann and J. Witkowski (New York, 2012).

6 Watson managed to damage his reputation as a nice man in other ways, from his commentaries on the general intellectual superiority of white people to the particular brilliance of James Watson; see Watson's biography at www.biography.com.

7 Pauling's triple helix was published as a note in *Nature*, followed by a detailed exposition in the *Proceedings of the National Academy of Sciences*, which shows the influence of his name and the editorial practice in the 1950s of accepting papers without additional review. See Melinda Baldwin, 'Credibility, Peer

Review, and *Nature*, 1945–1990', *Notes and Records of the Royal Society of London*, LXIX (2015), pp. 337–52.

8 S. Harding and D. Winzor, 'Obituary – James Michael Creeth, 1924–2010', *The Biochemist*, XXXII/2 (2010), available at www.biochemist.org.

9 If she had lived, Rosalind Franklin might have won a Nobel Prize for her X-ray studies on viruses, which she completed after her contributions to solving the structure of DNA.

10 Ralf Dahm, 'Friedrich Miescher and the Discovery of DNA', *Developmental Biology*, CCLXXVIII (2005), pp. 274–88. Pus cells are leucocytes, or white blood cells, produced by the immune system. Miescher chose them for his experiments because they were concentrated in the pus that collects on bandages around infected wounds, providing him with a pure source of a particular kind of cell. Miescher studied medicine in Basel and performed his research on pus cells in the medieval castle of Tübingen in southern Germany.

11 Georgina Ferry, *Dorothy Hodgkin: A Life* (London, 2014). Dorothy Hodgkin received the Nobel Prize in Chemistry in 1964. She is the only British woman to have received a Nobel Prize in any of the three science categories.

12 The gene *CFTR*, which is written in italics, encodes a protein, CFTR, which is written in roman type and is the acronym for the cystic fibrosis transmembrane conductance regulator.

13 Lindsey A. George et al., 'Hemophilia B Gene Therapy with a High-specific-activity Factor IX Variant', *New England Journal of Medicine*, CCCLXXVII (2017), pp. 2215–27; Savita Rangarajan et al., 'AAV5-factor VIII Gene Transfer in Severe Hemophilia A', *New England Journal of Medicine*, CCCLXXVII (2017), pp. 2519–30.

14 Watson, *The Annotated and Illustrated Double Helix*, p. 9, note 5.

NINE

GREENHOUSE

How We Make Things Worse

1 These opening sentences are adapted from Edward Gibbon, *The Decline and Fall of the Roman Empire*, vol. IV, Chapter 38 (New York, 1994), p. 119: 'The rise of a city, which swelled into an empire, may deserve, as a singular prodigy, the reflection of a philosophic mind. But the decline of Rome was the natural and inevitable effect of immoderate greatness. Prosperity ripened the principle of decay; the causes of destruction multiplied with the extent of conquest; and, as soon as time or accident had removed the artificial supports, the stupendous fabric yielded to the pressure of its own weight. The story of its ruin is simple and obvious; and, instead of inquiring *why* the Roman empire was destroyed, we should rather be surprised that it had subsisted so long.' If you can make time for the immersion necessary to read this six-volume masterpiece, Gibbon's voice will become a companion for life.

2 Further information on these circumstances can be found via the following resources: global warming: https://climate.nasa.gov; ocean acidification: www.whoi.edu/ocean-acidification and http://nas-sites.org/oceanacidification/; plastic pollution of the oceans: www.sciencemag.org/tags/plastic-pollution; air pollution: www.who.int/airpollution/en; deforestation: www.worldwildlife.org/threats/deforestation; grassland loss: Karl-Heinz Erb et al., 'Unexpectedly Large Impact of Forest Management and Grazing on Global Vegetation Biomass', *Nature*, DLIII (2018), pp. 73–6; shrinking lakes: Kate Ravilious, 'Many of the World's Lakes are Vanishing and Some May be

Gone Forever', *New Scientist* (4 March 2016), available at www.newscientist.com/article/2079562 (2016); desertification: www.un.org/en/events/desertificationday; soil erosion: Pasquale Borrelli et al., 'An Assessment of the Global Impact of 21st Century Land Use on Soil Erosion', *Nature Communications*, VIII/2013 (2017); population projections: www.un.org/development/desa/en/news/population.

3 Overview of threat to biodiversity from climate change: Rachel Warren et al., 'The Implications of the United Nations Paris Agreement on Climate Change for Globally Significant Biodiversity Areas', *Climatic Change*, CXLVII (2018), pp. 395–409; extreme weather: www.ucsusa.org; droughts: S. Mukherjeee, A. Mishra and K. E. Trenberth, 'Climate Change and Drought: A Perspective on Drought Indices', *Current Climate Change Reports*, IV (2018), pp. 145–63; large mammal extinctions: Felisa A. Smith et al., 'Body Size Downgrading of Mammals Over the Late Quaternary', *Science*, CCCLX (2018), pp. 310–13; loss of fisheries: Qi Ding et al., 'Estimation of Catch Losses Resulting from Overexploitation in the Global Marine Fisheries', *Acta Oceanologica Sinica*, XXXVI (2017), pp. 37–44; insect losses: Caspar A. Hallmann et al., 'More than 75 per cent Decline over 27 Years in Total Flying Insect Biomass in Protected Areas', *PLOS ONE*, XII/10 (2017), e0185809; plant losses: www.stateoftheworldsplants.com; microbial losses: S. D. Veresoglou, J. M. Halley and M. C. Rillig, 'Extinction Risk of Soil Biota', *Nature Communications*, VI/8862 (2015).

4 The NASA Global Climate Change website provides further information: https://climate.nasa.gov/vital-signs/sea-level; see also the IMBIE Team, 'Mass Balance of the Antarctic Ice Sheet from 1992 to 2017', *Nature*, DLVIII (2018), pp. 219–22.

5 Human origins have been complicated by evidence that modern humans emerged from multiple populations of *Homo sapiens* and matings with closely related species of *Homo*, which fanned out across Africa. See Eleanor M. L. Scerri et al., 'Did Our Species Evolve in Subdivided Populations across Africa, and Why Does it Matter?', *Trends in Ecology and Evolution*, XXXIII/8 (2018), pp. 582–94.

6 S. Wynes and K. A. Nicholas, 'The Climate Mitigation Gap: Education and Government Recommendations Miss the Most Effective Individual Actions', *Environmental Research Letters*, XII (2017), 074024.

7 Thomas Malthus, *An Essay on the Principle of Population* (London, 1798).

8 Paul R. Ehrlich, *The Population Bomb* (New York, 1968). The human population doubled in the half-century following the publication of this book. In 2009 Paul Ehrlich and his wife and co-author, Anne Ehrlich, wrote: 'Perhaps the most serious flaw in *The Bomb* was that it was much too optimistic about the future.' This verdict appeared in their essay 'The Population Bomb Revisited', *Electronic Journal of Sustainable Development*, I/3 (2009), p. 66.

9 Carbon dioxide levels also fell sharply during the Eocene Epoch, turning Earth from hothouse to icehouse. Marine organisms called diatoms, whose abundance rose in the Eocene oceans, may have been partly responsible for this transformation of the atmosphere. See David Lazarus et al., 'Cenozoic Planktonic Marine Diatom Diversity and Correlation to Climate Change', *PLOS ONE*, IX/1 (2014), e84857. These glass-shelled microbes absorb carbon dioxide and release oxygen, contributing as much to cooling and oxygenating the planet as rainforests on land.

10 Earliest butchery tools: Sonia Harmand et al., '3.3-million-year-old Stone Tools from Lomekwi 3, West Turkana, Kenya', *Nature*, DXXI (2015), pp. 310–15; hafted projectiles: Jayne Wilkins et al., 'Evidence for Early Hafted Hunting Technology', *Science*, CCCXXXVIII (2012), pp. 942–6; bow and arrow: Kyle S. Brown et al., 'An Early and Enduring Advanced Technology Originating 71,000 Years Ago in South Africa', *Nature*, CDXCI (2012), pp. 590–93.

11 Frédérik Saltré et al., 'Climate Change Not to Blame for Late Quaternary Megafauna Extinctions in Australia', *Nature Communications*, VII (2017), 10511.

12 R. P. Duncan, A. G. Boyer and T. M. Blackburn, 'Magnitude and Variation of Prehistoric Bird Extinctions in the Pacific', *Proceedings of the National Academy of Sciences*, CX (2013), pp. 6436–41; Morten E. Allentoft et al., 'Extinct New Zealand Megafauna Were Not in Decline before Human Colonization', *Proceedings of the National Academy of Sciences*, CXI (2014), pp. 4922–7.

13 Smith et al., 'Body Size Downgrading of Mammals', pp. 310–13.

14 Nancy L. Harris et al., 'Using Spatial Statistics to Identify Emerging Hot Spots of Forest Loss', *Environmental Research Letters*, XII (2017), 024012.

15 Quirin Schiermeier, 'Great Barrier Reef Saw Huge Losses from 2016 Heatwave', *Nature*, DLVI (2018), pp. 281–2; Terry P. Hughes et al., 'Global Warming Transforms Coral Reef Assemblages', *Nature*, DLVI (2018), pp. 492–6.

16 Hallmann et al., 'More than 75 per cent Decline over 27 Years'.

17 Listing for *Homo sapiens* on the IUCN Red List of Threatened Species, available at www.iucnredlist.org/details/136584/0.

TEN

GRACE

How We Should Leave

1 Jean-Daniel Collomb, 'The Ideology of Climate Change Denial in the United States', *European Journal of American Studies*, IX/1 (2014). This article reviews some of the ideological foundations of climate change denialism in the United States.

2 A. S. Mase, B. M. Gramig and L. S. Prokopy, 'Climate Change Beliefs, Risk Perceptions, and Adaptation Behavior among Midwestern U.S. Crop Farmers', *Climate Risk Management*, XV (2017), pp. 8–17; J. E. Doll, B. Petersen and C. Bode, 'Skeptical but Adapting: What Midwestern Farmers Say about Climate Change', *Weather, Climate and Society*, IX (2017), pp. 739–51.

3 B. Basso and J. T. Ritchie, 'Evapotranspiration in High-yielding Maize and Under Increased Vapor Pressure Deficit in the U.S. Midwest', *Agricultural and Environmental Research Letters*, III (2018), 170039. Other studies indicate consistently declining yields of maize, soybean and wheat under warmer conditions: Bernhard Schauberger et al., 'Consistent Negative Response of U.S. Crops to High Temperatures in Observations and Crop Models', *Nature Communications*, VIII (2018), 13931.

4 Tamma A. Carleton, 'Crop Damaging Temperatures Increase Suicide Rates in India', *Proceedings of the National Academy of Sciences*, CXIV (2017), pp. 8746–51. Carleton's study attracted some criticism and she defended her work in a detailed response: T. A. Carleton, 'Reply to Plewis, Murari et al., and Das: The Suicide-temperature Link in India and the Evidence of an Agricultural Channel are Robust', *Proceedings of the National Academy of Sciences*, CXV (2018), pp. e118–21. General concerns

about the impact of climate change on the mental health of children were raised by H. Majeed and J. Lee, 'The Impact of Climate Change on Youth Depression and Mental Health', *Lancet Planetary Health*, I (2017), e94–5.

5 A. Cunsolo and N. R. Ellis, 'Ecological Grief as a Mental Health Response to Climate Change-related Loss', *Nature Climate Change*, VIII (2018), pp. 275–81.

6 Maggie Astor, 'No Children Because of Climate Change? Some People Are Considering It', *New York Times* (5 February 2018); the quoted sentiment, which has a heartbreaking, Miltonic resonance, was made in an online post stimulated by an essay: Madeline Davies, 'With Environmental Disasters Looming, Many Are Choosing Childless Futures', 5 February 2018, www.jezebel.com.

7 Roy Scranton, *Learning to Die in the Anthropocene: Reflections on the End of Civilization* (San Francisco, CA, 2015), p. 21.

8 A. R. Jadad and M. W. Enkin, 'Does Humanity Need Palliative Care?', *European Journal of Palliative Care*, XXIV (2017), pp. 102–3.

9 Eric M. Jones, 'Where Is Everybody?', *Physics Today*, XXXVIII (1985), p. 11.

10 If you need a reminder, atom bombs work by nuclear fission and hydrogen bombs obtain additional explosive power by combining fission with fusion reactions. In an atom bomb, the detonation of a conventional chemical explosive forces radioactive atoms of uranium or plutonium together, causing them to split into lighter elements with the release of heat and gamma rays. Hydrogen bombs, or thermonuclear weapons, use this type of fission reaction to trigger a second fusion reaction that releases even more energy.

11 David C. Catling, *Astrobiology: A Very Short Introduction* (Oxford, 2013).

12 No disrespect is intended towards Pastafarians: see www.venganza.org.

13 Christiana Figueres et al., 'Three Years to Safeguard Our Climate', *Nature*, DXLVI (2107), pp. 593–5. The authors of this provocative commentary suggest goals for 2020 that would allow us to limit warming to 1.5°C increase above pre-industrial levels. This was the threshold set at the Paris Agreement in 2015.

14 The memorable Jim Morrison quote comes from his poem 'American Night', performed on the album by The Doors titled *An American Prayer* (Elektra/Asylum Records, 1978).

15 Arthur Schopenhauer, *Studies in Pessimism: A Series of Essays by Arthur Schopenhauer*, trans. T. B. Saunders (St Clair Shores, MI, 1970), p. 11, emphasis in original. In *Middlemarch*, George Eliot wrote about hearing the suffering in the world as 'a roar that lies on the other side of silence'.

16 The term 'biophilia' was coined by Erich Fromm in *The Anatomy of Human Destructiveness* (New York, 1973). E. O. Wilson popularized the idea in *Biophilia* (Cambridge, MA, 1984).

17 Ryan Gunderson, 'Erich Fromm's Ecological Messianism: The First Biophilia Hypothesis as Humanistic Social Theory', *Humanity and Society*, XXXVIII (2014), pp. 182–204.

18 Wilson attempted to get around this objection by broadening the definition of biophilia to include the option of an innate aversion to nature. This was as nonsensical as shoehorning the dislike of England and English people into the meaning of Anglophilia. A detailed critique of biophilia is offered by Y. Joye and A. de Block, '"Nature and I Are Two": A Critical

Examination of the Biophilia Hypothesis', *Environmental Values*, XX (2011), pp. 189–215.

19 Eric Jensen, 'Evaluating Children's Conservation Biology Learning at the Zoo', *Conservation Biology*, XXVIII (2014), pp. 1004–11; Michael Gross, 'Can Zoos Offer More Than Entertainment?', *Current Biology*, XXV (2015), pp. R391–4.

20 Samuel Beckett, *Waiting for Godot: A Tragicomedy in Two Acts* (New York, 1954), Act II, p. 61.

21 Aeschylus, *The Oresteia*, trans. Robert Fagles (London, 1984), p. 50.

ACKNOWLEDGEMENTS

I thank Diana Davis and Judith Money for reading the first drafts of the chapters and highlighting sections that required greater lucidity. Zack Hill, poet and screenwriter, served as my go-to grammarian and most helpful critic. Support from Michael Leaman at Reaktion Books has meant more to me than I have expressed via e-mail. He is among the least selfish of all apes.

INDEX